Information Spread in a Social Media Age

Information Spread in a Social Media Age

Modeling and Control

Michael Muhlmeyer
Shaurya Agarwal

CRC Press
Taylor & Francis Group
Boca Raton London New York

CRC Press is an imprint of the
Taylor & Francis Group, an **informa** business

MATLAB® is a registered trademark of The MathWorks, Inc. For product information, please contact: The MathWorks, Inc. 3 Apple Hill Drive Natick, MA 01760-2098 USA Tel: 508-647-7000 Fax: 508-647-7001 Email: info@mathworks.com Web: www.mathworks.com

First edition published 2021
by CRC Press
6000 Broken Sound Parkway NW, Suite 300, Boca Raton, FL 33487-2742

and by CRC Press
2 Park Square, Milton Park, Abingdon, Oxon, OX14 4RN

Library of Congress Cataloging-in-Publication Data

ISBN: 978-0-367-20871-4 (hbk)
ISBN: 978-0-367-71396-6 (pbk)
ISBN: 978-0-429-26384-2 (ebk)

To our families, our friends, and our colleagues. Without their help and support, this book would not be possible.

Foreword

In this current time of social and political upheaval, with the added challenge of a raging pandemic, this book, Information Spread in a Social Media Age: Modeling and Control, is both timely and appropriate from an engineering, scientific, and policy point of view. On a global scale, society is experiencing an unprecedented period of tremendous technical growth, leading to a never before seen and increasingly high level of connectivity between people and groups. While the global nature of a networked society has some great advantages, there are also some severe challenges. Misinformation and disinformation can be easily spread through our cyber-social networks creating dangerous consequences for humanity. Hence, it is imperative for the survival and prosperity of society that we understand the mechanics of this new networked structure to keep its benefits while eliminating and controlling any deleterious aspects. This book addresses this very specific topic in a seamless way covering not only the modeling aspect of the structure and dynamics of the networked systems but also control design, which relates to public policy and management issues.

The best part of the book is the way in which it presents the social and network theory, followed by various models of information spread, and then finally mechanisms for control design for the management of the dynamics of the system. The book is written in a gentle tone, making it easy to comprehend not only the models and control designs, but also the reasoning behind each choice that the modeler or a policy maker might make. The level of exposition included makes it easily accessible for readers from disparate backgrounds, such as social sciences, engineering, public policy, etc. The book also comes with detailed software code that shows simulations of the models and the performance of control designs. The reader can easily understand the software code and put it to use for their own purposes.

The book comes from a place of academic strength that relies on original research work performed by the authors of this book on this topic. Moreover, the book is presented in a way that makes it usable as a tool to learn modeling and control design basics and to code those models using software while keeping the main topic of social media networks and information spread maintained as examples.

The book can be fruitfully enjoyed by beginners in their respective fields. However, the book also provides a very serious look at cutting edge research in these topics and prepares readers to engage in these subjects themselves. The models and control design techniques can be modified to tackle other similar problems not only in social networks and information spread but other areas as such as studying the impact of COVID-19 and designing effective strategies for tackling it in a methodical way.

University of Nevada, Las Vegas   Pushkin Kachroo
Las Vegas, NV, USA

Preface

Approximately 3.5 billion people make use of social media worldwide, and usage is proliferating. That is nearly half of the entire global population. Social media is everywhere! From socially connecting on Facebook and Instagram to discussing news events with strangers on Reddit, it's hard to envision our modern world without a strong social media presence. Topics such as fake or unverified news stories, targeted online ads, and political polarization techniques are becoming important areas of discussion, research, and concern. In this text, we seek to explore these issues, not only from a general and high-level perspective but in technical detail, as well. Our goal in writing was to create a text of varying complexity to be enjoyable and informative to casual and professional readers alike who wish to explore the socio-technical aspects of social media and information propagation. Our primary focus is on the macroscopic modeling and design of control algorithms for these complex systems.

This book is targeted toward upper-level undergraduate and graduate students interested in topics such as information spread, macroscopic mathematical modeling, social science modeling, and control. While we primarily focus on social media-based examples and framing, the topics within this text can be useful for general applications as well. We hope that casual readers outside of these areas will also gain some appreciation and understanding of these topics. We provide fun and intuitive examples and real-life case studies from popular culture using social marketing and epidemiology-based information-spread models, including several novel information spread and social marketing models we developed and published throughout the last few years. Additionally, we provide `MATLAB` code for simulating many of our models to allow readers to explore further.

The book is organized into three main parts: (1) social networking in popular culture and basic social network theory; (2) macroscopic mathematical modeling as it ultimately applies to information spread; and (3) control methods to be applied to such models. We encourage those unfamiliar with the topics to read at least the first two or three chapters of each section to gain a decent fundamental understanding of the material. Advanced readers of specific interest areas may choose to skip the initial chapters and focus on the mathematical modeling of particular information spread scenarios, case studies, or control applications.

Why did this book come about in the first place? It was prompted not just by our interest in the topic, but also because we discovered the need for such a text during our research. What started as a university thesis between professor and student developed further into research papers and eventually into this text. Many models discussed in the book are original works of the authors. The proposed models and case studies have been published in peer reviewed IEEE journals. The content and presentation has been adapted from the previously published authors' research articles to suit this book. Given our engineering backgrounds, it was interesting to use our technical skills and apply them to a seemingly unrelated social field in new and exciting ways. We learned a great deal during this process, and developed a fresh and unique perspective on information spread.

We couldn't have done any of this without the input and expertise of our colleagues, and the support of our friends and families. They were all instrumental in the writing of our first book. Finally, we would like to acknowledge all those interested in information spread over social media. Your interest and curiosity in the topic is our inspiration to continue working for this book and beyond. We hope you enjoy the journey as much as we enjoyed marking out the path.

Los Angeles, CA, USA   Michael Muhlmeyer
Orlando, FL, USA   Shaurya Agarwal

Authors

Michael Muhlmeyer is an aerospace Communications Systems Engineer in Redondo Beach, CA. Previously, he served as a Senior Researcher at SmartCityLab in the Electrical and Computer Engineering Department at California State University, Los Angeles (CSULA). He received his M.S. in Electrical Engineering from California State University, Los Angeles. His bachelor's degree, also in Electrical Engineering, was obtained from California State University, Northridge (CSUN). He has authored two journal publications which have appeared in IEEE Transactions on Computational Social Systems, and IEEE Systems Journal. His areas of interest include mathematical modeling, control systems, computational social systems, information spread on social media, fake news, and novel applications of engineering to multidisciplinary research.

Shaurya Agarwal is currently an Assistant Professor in the Civil, Environmental, and Construction Engineering Department at the University of Central Florida. He was previously an Assistant Professor in Electrical and Computer Engineering Department at the California State University, Los Angeles. He completed his Post Doctoral research at New York University and Ph.D. in Electrical Engineering from University of Nevada, Las Vegas. His B.Tech. degree is in Electronics and Communication Engineering from Indian Institute of Technology (IIT), Guwahati. His research focuses on interdisciplinary areas of cyber-physical systems, smart and connected communities, and socio-technical-infrastructures systems. Passionate about cross-disciplinary research, he integrates control theory, information science, and mathematical modeling in his work. He has published over 23 peer-reviewed publications and multiple conference papers on various topics including information spread on social media, social media satisfaction, and intelligent transportation systems. His work has been funded by several private and government agencies.

Acknowledgments

The journey to write this book started around August 2016. This book is now finally complete after many delays and hitting many roadblocks. Authors Michael Muhlmeyer (MM) and Shaurya Agarwal (SA) would like to thank everyone who supported it. It is always challenging to acknowledge everyone for their support, but we will thank as many as possible.

Contributions from our colleague, Jiheng Huang, have enriched this book, like our joint research papers. The three of us have been involved in modeling social media information spread and its optimal control for several years. Thanks to his insight, dedication, and knowledge background, we could accomplish much more than we had ever hoped.

MM: I would like to specifically thank my amazing wife, Lili. From reading and helping me edit drafts to giving me advice on the cover to keeping me sane and encouraged throughout the entire process, she was as important to this book getting done as I was, and it would never have been possible without her love and support. I would also like to extend my sincere appreciation to my family. To my parents, Craig and Linda, I thank you for believing in me and guiding me through good and difficult times along my journey. To my siblings, Sabrina, Erik, and Nicholas, I thank you all for sustaining me in ways that I never knew that I needed.

SA: Life is a journey, and continuous learning is an integral part of that journey. I was fortunate enough to have found wonderful teachers throughout who have inspired me and instilled in me the value of education and hard work. I want to thank my lifelong gurus — Dr. Mohini Goel (mother and the first teacher), Mrs. Inez Smith, Late Shri. Ugrasen Singh, Mr. T.N. Mishra, Dr. Pushkin Kachroo — and all my other teachers who have helped me reach where I am today. I also thank my father — Mr. Sanjay Agarwal, and brother — Shivam, for their continuous support and encouragement. A special thanks to my lovely wife, Saumya, for standing by me through the thick and thin. Her unrelenting support and constant motivation kept me going during the tough days of my Ph.D. and later, during the writing of this book. Last but not least, to my daughter, Shanya, with whom I am growing up again, exploring new dimensions every day! I am also thankful for the support I received at the California State University Los Angeles and the University of Central Florida.

Thanks to everyone on the Taylor and Francis team who helped us, particularly during the difficulties surrounding the COVID-19 crisis and other setbacks. Special thanks to Nora, the ever-patient Editorial Director, and

Prachi, our Editorial Assistant. You were both a pleasure to work with and a great help in providing the resources and expertise needed to turn our research and work from a disorganized set of papers into a professional book publication. Authors also thank Dr. Pushkin Kachroo for writing a foreword of this book and providing insightful comments.

Finally, we would like to express our profound gratefulness for the insightful and thought-provoking interactions during our academic and professional careers with undergraduate and graduate students, colleagues, collaborators, and fellow researchers.

List of Figures

List of Tables

List of Codes

Symbols

Symbol Description

N	Population size or number of nodes in a network.
v	Node of interest.
C_D	Degree centrality of a network.
C	Closeness centrality.
d	Distance between x and y vertices.
t	Time.
$\mathbf{x}$	State vector.
y	Output vector.
$\mathbf{u}$	Control vector.
$\mathbf{A}$	System matrix in state space representation.
$\mathbf{B}$	Input matrix in state space representation.
$\mathbf{C}$	Output matrix in state space representation.
$\mathbf{D}$	Feedback or feed-forward matrix in state space representation.
I	Ignorant class for information spread.
S	Spreading class for information spread.
R	Recovered class for information spread.
C	Counter-spreading class for information spread.
E	Exposed class for epidemic spread.
β	Spreading rate.
γ	Stifling rate.
δ	Decay rate in information interest.
α	Counter-spreading rate.
μ	Counter-stifling rate.
ω	Willingness of spreaders to listen to counter-spreaders (receptivity rate).
σ	Stochastic element of a system ($\sqrt{b(S) + d(S)}$).
k	Average connectedness factor.
d	Strength of information influence factor.
R_0	Basic reproduction number: Average number of secondary infections from a single infected individual ($\frac{\beta}{\gamma}$).
$\mathbf{W}$	Wiener process.
$\mathbf{K}$	Control gain.
$\mathbf{L}$	Luenberger observer gain.
e	Error function.
$\mathbf{J}$	Performance measure.
$\mathcal{H}$	Hamiltonian.
$\mathbf{u^*}$	Optimal control.
λ	Time to be optimized.

1

Introduction

> *Haven't you noticed that opinion without knowledge is always a poor thing? At the best it is blind—isn't anyone who holds a true opinion without understanding like a blind man on the right road?*
>
> Plato, *The Republic*, 375 BC

Socio-technical systems, or systems that intersect societal and technological domains, are becoming commonplace in our everyday lives. This is especially true with the rise and widespread acceptance of online social media applications and websites. Thanks to cell phones, laptops, smart televisions, and easily accessible high-speed internet, social media in its many forms have become one of our primary sources of information and cultural exchange. As a result, most of us have grown accustomed to the usage of social media terminology and concepts in our everyday lives, some of which are shown in Figure 1.1.

FIGURE 1.1: Social media terminology.

Information is propagated via these social media networks throughout the internet, forming or changing opinions, beliefs, and understanding. It's hard to understate our increasing level of connectivity to social media as it begins to dominate many aspects of our communication, news acquisition, cultural consumption, socialization, and professional lives.

Given the ubiquity of social media and the need to explore, understand, and perhaps influence it, this book focuses on information propagation as it applies to social media networks. Specifically, the focus is on the macroscopic modeling and control of information spread in these complex socio-technical systems.

In this chapter, the overarching concept of information spread is introduced and framed in the specific context of information spread over online social media. It begins with a loose classification of the types of information examined in this text along with the reasons behind an interest in understanding and analyzing these types of information. Several modern scenarios focusing on information spread are presented as real-world examples including mass communication, governmental concerns, advertising, political campaigning, misinformation, and more. The concept of potentially controlling the spread of information is also addressed, and will be expanded upon in future chapters. Finally, recommendations on how to read the text are presented based on the background of the reader, ranging from covering fundamental principles in accessible language and examples to delving into more advanced modeling, simulation, and control techniques.

1.1 Expressions of Information

Information for this text is broken down into three main expression categories: factual information, "opinion-based" information, and contentious information, as shown in Figure 1.2.

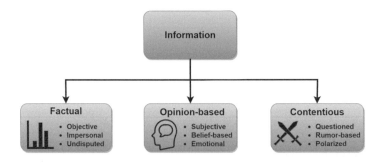

FIGURE 1.2: Common expressions of social media information.

Factual information is essentially raw data and truths. It includes mathematics, statistics, natural disasters, a documented event, and similar undisputed, impersonal, and nondiscriminatory information. To paraphrase Plato, while opinions are bound to change over time, facts are different. They remain unchanged, or at best, get added to or modified. Mathematical facts are good examples of factual information. The existence of numbers can be considered a fact. Although we can philosophically dispute whether numbers really exist or are merely a necessary abstraction, such disputes at best refine our concept of numbers, not dissolve it. Questioning factual information does not dissolve it, only refine or add to it. We can trust facts to stand on their own.

Another example of factual information is consumer research habits. Conscientious consumers pay careful attention to accurate information out of the need to understand services or goods that are being sold, especially if it is advertised and purchased without being physically seen first [1]. Thus, when we shop with online retailers (for a book, for example), we are usually attentive to the factual information associated with a particular product, including size, dimensions, technical specifications, and more.

Factual information is a good candidate to be modeled using well-established mathematical epidemiology-based and Maki-Thompson models. These models are described in detail in the second part of this text, but can be briefly summarized as follows: there is an initial message or communication; the word is spread throughout the population; the message eventually reaches a saturation point where a certain percentage of people are aware of it. This form of information flow has been widely studied.

Opinion-based information is, for our purposes, "factual information" that results in a divided message between one or more groups. Each of these groups will interpret the qualitative portions of an otherwise fact-based message in a different way [2]. Divisions can occur over arguments involving unverified scientific discoveries, policies, elected officials, advertisements, entertainment media, and much more. Peoples' opinions involving a single topic can diverge into several factions. Still, each of these factions represents a large cluster of people sharing the same opinion and feeding validation of that opinion back into their faction group.

We can see several examples of opinion-based information in areas such as finance, where the differing opinions of stock traders are associated with larger trade volumes in the stock market [3]. Continuing with a previous example, that a book purchased from an online retailer has three hundred pages is factual information. A reader's review discussing whether or not the book was good or bad during the three-hundred-page read would be opinion-based.

In some ways, we live in a "post-truth" age of information spread. People are increasingly appealing to emotional divisions and personal beliefs over merely looking for and accepting factual or objective information [4]. As such, exclusive models are needed to observe and analyze this type of information. For opinion-based information analysis, two or more main groups with differing beliefs must exist. Additionally, some form of information exchange between them to spread opposing viewpoints from each group's core (otherwise isolated) opinions must also be present.

Contentious information is information where the accuracy, authenticity, or validity of allegedly "factual information" itself is questioned. Like opinion-based information, there is still back and forth (sometimes heated) debate over it. Here, however, knowledge itself is the subject of contention, not an interpretation of it.

There are many examples of contentious information in our everyday lives. Is organic or GMO (genetically modified organism) food safer? Did a beloved celebrity actually die, or was it only a social media rumor? Is this internet image authentic, or was it doctored before being distributed? These questions, as well as disputed "fake news", quotes, and out of context statistics, are all potentially factually information that can be viewed as contentious [5].

In the running example of online book purchases, the factual information is the book's three hundred page length. If this information was contentious, some would argue that the page count is not correct. Perhaps the online vendor is counting title pages, a table of contents, or an index, and there is contention as to whether those should "count" as content pages. At times, even scientific paradigms and discoveries can be distrusted and viewed as contentious by the general population, where expert claims and advice are openly disputed due to their inherent uncertainties [6].

There is not much research done in this area. Still, it is becoming a topic of increasing interest as social media increasingly connects information and beliefs between socially, culturally, and politically diverse population segments. As with other types of information spread, exclusive models will be needed to understand better, analyze, and possibly influence this type of information expression. However, research and modeling in this area have proven difficult due to data collection difficulties, privacy concerns, and differences in the backgrounds of people whose data are collected [7].

A final note must be made concerning the concept of misinformation, which is in opposition to facts, opinions, and contentious expressions made in good faith. Broadly, misinformation can be defined as information that is incorrect, either intentionally or by accident [8]. Sometimes misinformation is spread intentionally for an ulterior motive. Sometimes it is merely a mistake of getting the facts wrong or spreading information without prior verification. While fact-based information cannot be categorized as misinformation, both opinion-based and contentious information expressions can arise in the form of misinformation. The only difference is that these opinions and contentious debates would be based on underlying false information.

1.2 Why Information Spread Matters?

Humans communicate. We share ideas, technology, opinions, strategies, religion, art, and other cultural elements. In fact, spreading information is a large part of what separates humans from the vast majority of the animal kingdom. As such, information spread has always been an underlying part of not only the individual human experience but also our progress as a species. Until relatively recently, information spread (or "rumor" spread) was at best examined through intuition, experience, and assumptions. A disease outbreak in an early settlement could be communicated to neighboring communities through a messenger. Still, there was no way of knowing how far and to what degree that information would penetrate a targeted population. As writing became widespread, so too did the capacity of a community to spread the news within itself and to its neighbors. Later, more advanced technologies such as the printing press, telegram, telephone, radio, and television added to the older information spread methods, allowing a small number of individuals to spread information or communicate to the masses with relative ease. Finally, in the modern information age, high-speed near-instant communication is available to nearly every individual in developed societies.

Why does our ability to spread information matter? Because societies are profoundly affected by the quality and reach of information spread. The spread of a political candidate's message to as many voters as possible is what allows election campaigns to succeed. Products that are widely advertised result in higher consumer spending and greater corporate profits than those without such advantages. Precautions and information about natural disasters or other emergencies can be spread to neighboring areas to mitigate the effects of the catastrophe. Misinformation over social media and news sites can lead to poor individual choices and actions, possibly harming one's health and well-being. The examples are countless. As human beings living in a modern society, we rely on communication methods and information spread to function. Several industries and social structures can benefit from examining and analyzing the methods by which different types of information disseminate and recede. With advanced knowledge in areas such as sociology, mathematics, engineering, and information science, there are more tools available now than ever to effectively understand, predict, and control information as it moves throughout a targeted or generalized community.

1.3 Modern Information Spread Scenarios

We now discuss a few examples of social theory applications and recent information spread scenarios that further demonstrate the importance of understanding how it spreads as our technology, data acquisition capabilities, and communication methods advance.

1.3.1 Global Communication During a Pandemic

In March 2020, the World Health Organization (WHO) declared COVID-19 (coronavirus disease 2019) a global pandemic. COVID-19 infections resulted in a wide range of outcomes. Many infected individuals were asymptomatic, in which a person never showed negative symptoms and simply acted as a carrier of the disease. In more extreme cases, virus infection led to severe respiratory illness, and even death. Experts determined that the mitigation of the virus spread was critical and that certain general safety precautions should be taken by the general population, including mask usage, social distancing, and frequent hand sanitation [9].

On both traditional news outlets and social media, information began to spread rapidly throughout both the global and localized communities with the latest news, recommendations, government orders, infection rates, scientific research, and much more. While much of the information spread was fact-based, there was also opinion-based and contentious information. Was it right to force social distancing, mask requirements, and business closures, or do such actions infringe on individual freedoms? Can the data be trusted, or was it being used for political gains or social control? Social media discourse was awash with these and similar debates and discussions. Additionally, several instances of misinformation could be found throughout the internet, from possible harmless cures to rumors of supply shortages to dangerous unfounded medical advice. Misinformation propagation became so out of hand that official health agencies had to caution against rumors and provided steps for the public to help control coronavirus rumors [10].

1.3.2 Governments and Mass Panic

Consider the "Salt Panic" in China. In March 2011, a tsunami following the Tohoku earthquake led to three nuclear meltdowns, explosions, and the release of radioactive material in Japan. Upon hearing the news of the disaster along with a false rumor that iodized salt could help prevent radiation poisoning, panicked shoppers stripped Beijing stores of salt. As a result, salt prices were said to have increased up to tenfold in some areas. The Chinese government and international scientists repeatedly announced that there was no reasonable threat from the radiation. Even in the event of dangerous radiation reaching China, the essential table salt found in stores would not help mitigate any radiation effects. With efforts from local governments, the false rumors were eventually quelled. Regardless, the Salt Panic demonstrates the power of mass information spread, be it real or fictitious information [11].

1.3.3 Shopping and Advertising

Consider another example. Amazon, the popular online shopping site, has progressively gained better and better insight into patron purchase habits. By tracking what customers view and have purchased the past, they offer targeted recommendations. In addition to this, many users actively post reviews of products and seek out reviews for potential products they are considering for purchase. While the first element is simply accomplished via machine algorithm and data acquisition, the second is a direct result of active information spread within an online shopping community. While advertisements are designed to entice customers to buy a product, reviewers may give positive or negative feedback and ratings, which could have widespread influence over the general positive or negative perception and value of the product. Online word of mouth is an integral part of the modern online shopping experience, which cannot be ignored.

1.3.4 Social or Political Campaigning

The concept of campaigning is of particular interest in the study and application of information spread. In a campaign, measures are taken to spread a message throughout a population deliberately. These campaigns are often seen in the form of advertising campaigns for products or services and political campaigns for candidates or propositions. Effective spreading of the message can be especially powerful in these cases as the initial spreader is both creating information (accurate or not) and pouring resources into spreading it, hoping for enough traction and widespread belief. For advertisers, this means the product or service becomes popular, sells well, and brings financial benefits. For politicians, voters become aware of the candidate's highlighted promises, ideology, and qualifications (or negative attributes of political rivals) to ultimately gain votes. Both product advertisers and politicians pour vast amounts of capital into campaigns for a reason. It works and is, in fact, believed to be required for large-scale success over competitors.

Just as effective as building up a candidate or product, information spread can be utilized to tear down opponents. Countless smear campaigns are riddled throughout history. One need not look further than old election attacks by founders of the United States, Thomas Jefferson and John Adams. During election campaigning, Jefferson's hired attacker accused President Adams of having a *"hideous hermaphroditical character, which has neither the force and firmness of a man nor the gentleness and sensibility of a woman."* Adams' men called Vice President Jefferson *"a mean-spirited, low-lived fellow, the son of a half-breed Indian squaw, sired by a Virginia mulatto father."* Adams was labeled as a fool, a hypocrite, a criminal, and a tyrant, while Jefferson was branded as a weakling, an atheist, a libertine, and a coward [12]. The idea of actively spreading disinformation proved to be incredibly effective in democratic politics. Due in no small part to Jefferson's hired "hatchet man", he was able to win the first hotly contested Presidential election in the United States. Today, similar campaign attacks are frequent, oftentimes using exaggerated, out-of-context, or objectively false information.

Among famous product advertising campaigns, many may recall the popular *"Get a Mac"* television and internet campaign by Apple in 2006, in which personifications of Macintosh and Windows PCs introduce themselves as "I'm a Mac" and "I'm a PC" and proceed to act out various skits aimed at touting the benefits of a Macintosh over a Windows PC. The campaign was massively successful and gained popularity and recognition worldwide, leading to a 39% increase in Macintosh computer sales that year. Microsoft eventually released similar ads meant to parody and similarly appear superior to their competitors with nominal success [13].

1.3.5   Misinformation, Disinformation, and Fake News

Terms such as *misinformation, disinformation,* and *fake news* often confuse casual readers and researchers alike. This is because there is a little agreed-upon distinction drawn between these terms in modern research concerning these topics [14]. In general, however, we can broadly define misinformation as information that is simply incorrect, possibly by way of an accident or lack of knowledge. Disinformation, on the other hand, is often used to specify a subcategory of misinformation that is intentionally false [8].

The concept of "fake news" has recently become a hot topic in sociopolit-ical discussion, particularly during and immediately following the 2016 U.S. Presidential Election. That said, spreading a fake story or lie to further one's cause is hardly new. Let us revisit the 1800 United States Presidential Election. Adams lost the election not only because of the effectiveness of the smear campaigning but also because of the application of fake news via a deliberately false story that Adams wanted to go to war with France [12]. Similar examples of untrue stories framed as news can be seen throughout the United States and world history, mainly when the populace's views and opinions are essential (as they are in many democratically governed nations).

One must take care in differentiating between fake news and merely untrue rumors. Rumors can be real or false (and another form of misinformation), but they are generally a result of uninformed information spread. For instance, "hearing" that a celebrity has died and communicating it further (while the celebrity is, in fact, alive and well) may be inappropriate. Still, it is not being framed and presented in such a way as to be viewed by a typical reader as an official and verified true news story. Similarly, tabloid magazines with unverified stories are rarely perceived as reliable news sources and, in effect, are collections of (true or untrue) rumors. For there to be actual fake news, the story must be knowingly false to the initial spreader and framed in such a way as legitimate news.

Fake news can take many forms, from independent internet news sites to Facebook fan pages to WhatsApp sponsored messages [15]. While fake news can originate from any number of sources, it solidifies itself as appearing to be legitimate news once the story is picked up (and likely never properly fact-checked) by a widely read news source. At this point, the fake news can spread quickly and effectively in ways similar to legitimate news.

One major challenge currently is how to quickly identify and mitigate the proliferation of misinformation, disinformation, and fake news. With the advent of social media as a source of accepted news, both verified and fake news spread quickly on social networks (which can mean global spread in some instances) with little to no vetting. In fact, in modern news cycles, there is enormous pressure to release news as fast as possible, often circumventing traditional news journalism fact-checking procedures. As a result, fake news is more easily absorbed into and spread throughout the public consciousness as legitimate news, true or not. Thus, new strategies must be developed to either mitigate or proliferate fake news (depending on one's goals) to keep pace with the modern digital age of information spread.

1.4 Controllable Information Spread

Understanding modern applications of social media information spread is both academically and practically beneficial, but what can be done with this understanding? Typically, once a good qualitative and quantitative understanding of a domain is achieved, the natural response is to attempt to influence or control that domain. The term "control" can have several negative connotations in the context of social systems, but here we mean it in a neutral and academic sense. Control is simply an active way of influencing an object, system, event, or domain to yield desired results.

How then does one control information spread? Can such control be effective with the complexity of opinions and beliefs found in an internet population? The answer is somewhat complicated, but in short, yes. Through advertising, biased news stories, campaigning, and general social influencing, overall trends can emerge that sway public and individual opinions and beliefs. Is it precise? Not really, but it is influence and control nonetheless and still proves generally effective and useful. There's a good reason why vast sums of money are gladly spent on campaigning and product marketing: it works. One current goal of information propagation research is to quantify how to have information spread evolve with greater precision and to apply more nuanced control methods to a context-specific instance.

1.5   How to Read This Book

This book is divided into three main parts: social network theory, mathematical modeling, and control. While many readers will benefit most from reading each part and chapter in the order presented, we encourage those with existing knowledge of some topics to skip to different chapters as desired. If casual readers find that concepts are becoming too in-depth for their respective interest or proficiency levels, feel free to move to a new chapter where new concepts are first presented in high-level and easily accessible detail. Figure 1.3 gives a visual summary of the part and chapter organization of this book.

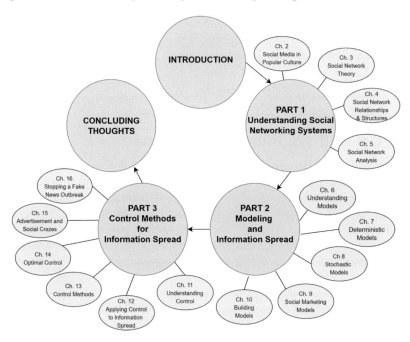

FIGURE 1.3: Overview of the organization of this book.

Part I explores the basics of social network theory and presents several relationships, definitions, and structural forms for networks, with a focus on social networking. Part II delves into the concept of models, beginning with general modeling theory and development steps and ending with several models presented to describe different types of information spread. Finally, Part III gives an overview and basic tutorial for fundamental control theory and their application in information propagation.

1.6 Exercises

1. Depending on the context, there are many ways information can be categorized. Systematically build a few of such categories and justify your reasoning.

2. Discuss, with examples, the differences between — *misinformation*, *fake news*, and a *rumor*.

3. Provide an example of a fake news story that you experienced on social media. How did you realize that it was indeed fake?

4. In this chapter, several reasons are given to justify the importance of information spread. List and justify at least three additional examples of how knowledge of information (or misinformation) spread can be useful either as an individual or a community.

Part I

Understanding Social Networking Systems

2

Social Media in Popular Culture

> *If you want to go fast, go alone. If you want to go far, go together.*
>
> African Proverb

Social media is a collection of websites, applications, services, and other computer-based mediums that allow people to share information and communicate. The concept of "connections" is key to social media. These connections can include friends, colleagues, people with similar interests, and more. These connections ultimately lead to virtual communities that simulate traditional community groups in varied areas as fan clubs, friend reunions, and art showcases. Thanks to main hubs (usually search engines), such as Google, Bing, and Yahoo, social media category groups can interact and influence one another.

This chapter highlights the role of online social media in popular culture. It begins with a visual and explanatory high-level topology of online social media. Some of the most popular social networking sites are presented and generally described for the benefit of the reader including Twitter, Facebook, and LinkedIn. Other branches of social media are also explored, such as content sharing sites, discussion forums, news and blog sites, online shopping and review platforms, and social media revolving around games and music. Concrete examples of well-known platforms and websites that fall under each social media branch are presented. Finally, some platforms and social media phenomena such as internet memes are expressed as a hybridization of other social media categories, as they fulfill multiple usage functions of generalized social media topology groups simultaneously.

2.1 The Topology of Social Media

How might we begin to organize and define the vast breadth and depth of internet social media sites? Unfortunately, there is no explicit way to classify every significant social media community, mainly since there often exists a considerable overlap between any two or more categories. However, a general (but not explicit), social media topology is given in Figure 2.1. In the next sections, we will examine some of these community types in more detail.

FIGURE 2.1: Topology of major social media sites.

2.2   Social Networking Sites

When examining social media in a practical and modern sense, a discussion will typically be within the context of popular digital social networks that are used to absorb and spread information within a population. More than any other type of social network in the past, digital social networks have revolutionized the speed and reach of information spread. News, rumors, and advertisements can reach from one section of the globe to another in mere seconds indirectly through these networks. These digital (or online) social networks are specifically designed to collect and form online communities and encourage the spread of information within the group and between adjacent groups.

2.2.1   Twitter

Twitter is a social networking site that allows users to post a short limited-character message over the internet via the Twitter website, a dedicated application, or a mobile device such as a cellphone. Twitter posters, or "Tweeters", will often post a "tweet" concerning what they are doing or thinking. The tweet is often accompanied by a reference tag known as a hashtag. It allows users to view similarly tagged tweets as a collection, usually referencing the same topic, event, or idea. Twitter is also used to post pictures, news, and current events. Many view Twitter as a quick and easy way to discover what is happening around the world by searching relevant keywords for news, trends, or current events.

Twitter is an especially popular social media platform for analyzing and collecting social network data due to its direct and traceable nature. Hashtag trends and "retweets" are relatively easy to manage and visualize compared to other online social media systems. Often, data science and machine learning algorithms for social research use Twitter as a primary data collection source.

2.2.2   Facebook

In contrast to Twitter, Facebook, as a social media networking site, focuses on each user's "News Feed". The News Feed is a page customized for each user, highlighting and tracking the activities of their fellow community "friends". Users make posts and friends of users can comment on or "like" said posts as a sign of agreement or interest. The idea of photograph or image sharing is much more pronounced and integral to Facebook than Twitter. On Facebook, each user's homepage is a collection of posts, discussions, and events based on the user's friend community's activities. Given the more personal nature of Facebook, there have been several concerns over privacy issues as to where and how this posting, liking, photograph viewing, and commenting information is shared.

Facebook is often marketed as a way for real-life friends to stay in touch after separations due to distance, change of lifestyle, or other key dividing factors that would otherwise cause individuals to lose track of one another slowly. Due to Facebook's personal nature, it has become a breeding ground for several forms of information spread, such as targeted advertisements, social movement growth, and news article distribution. Recently, Facebook has seen negative attention for a lack of content moderation, leading to concerns over the spreading of hate speech and the proliferation of fake news [16].

2.2.3   LinkedIn

While Twitter and Facebook are online social media networking sites catered toward news, opinionated commentary, and socializing, LinkedIn serves individuals and communities interested in employment-centered and professional networking. On LinkedIn, employers post job openings and company information pages, while job seekers set up professional profiles that include resumes, job experience, and curriculum vitae. Both employers and job seekers form "connections" through the site to build a network of like-minded professionals to pair employers with prospective employees. Users can follow various companies, "endorse" another user for a particular profile stated skill, post job listings, and more. Unlike many of the other social media sites, LinkedIn is not primarily concerned with casual information spread, so much as an advertisement (of one's self or one's company in this case).

Due mainly to its narrowly defined nature, LinkedIn is not seen as controversial or particularly interesting insofar as large-scale or viral information spread is concerned. Still, it is an excellent example of a small scale focused advertising network. Additionally, it is mostly free from some of the complexities of more open online social networks, such as fake news, political agendas, and socio-cultural divisions. Most users are simply there to post their employment information and have it spread sufficiently to find a connection with a suitable employer toward the end of gaining a job.

2.3   Content Sharing Sites

Many social media sites are dedicated to spreading information via media content distribution. This content can be created and uploaded to the site by a user or simply shared or linked by users who view the content and wish to spread it to others within the community. Usually, these sites have short comment sections for community members and use both content category tags and a "like" and "dislike" (or equivalent) system in order to feature media on the site's front page or target individual users based on interest. Despite the overarching theme of social sharing, sites go about expressing that sharing in different ways. For example, YouTube focuses on sharing videos; Instagram primarily shares pictures and short videos; GIPHY allows users to search for and share short looping videos with no sound resembling GIFs.

Typically, sites primarily focused on social content sharing will encourage users to re-post, link, or otherwise share existing media from other parts of the internet (accompanied by some comments). Savvy artists and content producers will use social sharing sites to display their artwork, videos, and comic strips to a broad audience with similar media interests.

2.4 Discussion Forums

Discussion forums are divided into a variety of topics, and users engage in "discussion threads" under a topic category. Each discussion beings with an original poster who will pose a question, make a statement, or generally post something that attempts to begin a group conversation. Users see these posts and reply based on their opinions, knowledge, or feelings on the topic. In political forums, for example, the original post might be a negative meme about a political candidate, with replies encouraging or disparaging the meme's general sentiment. Game-based discussion forums may be more focused on an optimal game strategy over subjective opinions about the game's enjoyment value. Local or event-based forums will be targeted to a specific community and discuss local events, community concerns, and nearby businesses.

Popular discussion forums include Reddit, Quora, Gamespot, and Stack Overflow. Topics of these discussions can consist of general issues, answers to specific questions, video game culture, and programming techniques, to name a few. Given the prevalence of comments in other forms of social media as well as posting pictures and videos in dedicated discussion forums, the lines of what constitutes a "discussion forum" can become blurred with other categories. Generally, however, the site or service should primarily cater to a discussion over other media in this category.

2.5 News and Blogs

News sites are widespread and straightforward to conceptualize. Generally speaking, many trusted news sites are simply online counterparts of a news source that has a newspaper or television presence. Sometimes, online-only news sources exist, particularly for niche topics such as news that highlights the latest tech gadgets, or an online news site that collects and distributes upcoming movie spoilers from inside sources. To be considered a legitimate news source, the organization and writers must be trusted to be reasonably impartial and put a reasonable effort into fact-checking news before it is published online. Many sites try to pose as legitimate news, but in reality, they have a hidden agenda or interest, which causes it to report unverified or even false stories and paint them as genuine. This practice is what is normally referred to as "fake news" or active misinformation.

Blogs, short for "weblogs", are primarily internet-based informal opinions and discussions on a specific topic. Their main purpose is to express an opinion or viewpoint and oftentimes to encourage discussion in a comments section following the blog article. It is important to note that blogs do not attempt to present themselves as official news, however well written and thoughtful they may be. Still, they are an essential medium for spreading information and especially opinions within a population. Especially popular blogs might become linked to an online social media site and spread far beyond their initial intended audience. While most blogs are text-based, they can also be comprised primarily of video, artwork, photographs, or audio (in the case of podcasts).

2.6 Shopping and Reviews

For many of us, shopping via social media is an integrated (but perhaps invisible) part of our everyday lives. We search for nearby local businesses on Google Maps. We buy something on Amazon that we need, but don't want to go from store to store searching for the perfect item. These kinds of consumer-based social media groups include shopping networks and trade-based networks. Targeted advertisements and brand or site loyalty play an important role in this social media category.

Many online shopping sites encourage postings in the form of user reviews. A consumer review subnetwork critiques multiple brands, products, businesses, or services and is often attached to a shopping or travel network. Reviews are becoming increasingly important for a business to thrive, as the information spread of their positive or negative traits will influence users to neglect or frequent a business. The businesses affected by these reviews strive to spread the information of good reviews and mitigate bad reviews.

2.7   Games and Music

Game and music culture is an important and growing segment of social media on the internet. Children, teenagers, and young adults frequently watch videos and read strategy guides for video and tabletop games. Music lovers rely on music curators to help keep up with the latest artists and trends. Within specific music and gaming communities, there are forums, video sharing, and imaging sharing mechanisms to feed an overall media community for any particular interest.

Games hold a special place in social media. With the rise of user-to-user connectivity, two users can play a game remotely despite not being in the same house, or even the same country. Sometimes a large group of users can play a shared game simultaneously. As a result, players interact inside of a game and share their in-game strategies, knowledge, beliefs, and ideas with other players. Game strategies evolve as users grow in skill, in-game friendships or clans grow, and biases against certain players or play-styles emerge.

While games and music are large industries with millions of users and wide acceptance in popular culture, more niche interests can have similar categorizations, each with their discussion forums, instructional videos, and inside jokes. Games and music, however, are two examples of hobbies that have evolved to become part of the public consciousness.

2.8   Hybrid Social Media

Clearly, every social media site or application will not fit neatly into the categories presented here (or perhaps any category). Let's examine Tumblr as an example. It was designed as a social media sharing application and website. Users post pictures, GIFs, and clips of other shared content and give brief tags or comments on that particular piece of media. Over time, Tumblr evolved to include many full-sized blogs, stories, fan-art, and fan-fiction.

Reddit, one of the most popular discussion forums, commonly includes shared links to internet memes, artwork, shopping deals, and news. Due to the nature of discussions, users will naturally reference other websites and media. While Reddit is still primarily a discussion forum, it will overlap with other social media sites with wildly different purposes in order to further or continue an online discussion. Everything from world news to shared jokes to video game deals will often cross the line between social media communities. As social media sites grow, develop, die, and consolidate, their typical usage will likewise change. The presented categories will move or overlap as users continue to define how a service will be utilized.

FIGURE 2.2: Example of an internet meme.

2.8.1 Internet Memes

Another hybrid form of social media is an internet meme. Memes are typically, but not necessarily a comedic piece of media (such as a video, image, hashtag, phrase, etc.) that spreads from one individual to another via the internet. Memes are often cultural symbols and social ideas that have a tendency to spread in a viral manner. The emergence of meme usage (along with emojis and GIFs) is often attributed to communication becoming increasingly visually dominant in order to save time or better express ideas beyond those expressible with words alone.

One enduring example of an internet meme is the "Rickrolling" prank. People send a seemingly legitimate internet link, which leads instead to a Rick Astley music video from the 1980s. Another good example is the popular trend of using photo editing software to alter movie posters in extreme and comical ways to express a point and post the new image on social media sites. Many memes that become widespread evolve to suit new situations and desired subjects of commentary. During the 2020 coronavirus pandemic discussed earlier, for example, people were given "stay at home" orders for several months. Many took the time to learn to bake bread since they were unable to go about their routine (generally more exciting) plans for the year. Several memes similar to the one shown in Figure 2.2 were created and posted over social media sites to express frustration through humor.

Internet memes are an excellent way to examine information spread because they leave behind a digital footprint of where they have been on social media and the internet at large. While some may die out and others become viral, they are reasonably easy to trace compared to other types of information. Additionally, because they are usually satire and not viewed as contentious information (memes are either shared or ignored), studying how internet memes spread simplifies many difficult and unpredictable elements encountered when studying other types of information propagation.

2.9 Exercises

1. What social media sites do you use? Using the taxonomy presented in Figure 2.1 map the categories of the sites you use.

2. Select at least four social media sites or applications not already presented. How would they be categorized? Why? Could they fit into more than one category?

3. Discuss how social media in popular culture has changed the information dissemination and propagation in society.

4. Building on the previous question, discuss how social media outlets have potentially changed society in a positive way.

5. Find an internet meme. Explain the message it is trying to convey. Where did the image or idea originate, and how does it relate to the meme?

6. Do you think there is a rise in misinformation or false news on several social media outlets? How does this affect your social media use and society in general?

7. Building on the previous question, what might be some possible options to control the spread of misinformation on the internet?

3

Social Theory and Networks

> *If you wish to win a man over to your ideas, first make him your friend.*
>
> Abraham Lincoln

Due to the breadth of areas of study required to have a fundamental understanding of mathematical modeling and control of online social media information spread, a summary of some previous works is helpful. As such, social theory is addressed in this chapter, particularly how social theory applies to social networks. While relatively brief, it lays the fundamental groundwork for understanding information spread in more breadth. The chapter begins with a condensed history of social theory and the philosophical thought that led to its development from the ancient world to the contemporary world. It moves on to modern social theory concepts and some of the theory and academic literature surrounding social networks. The social exchange theory is specifically examined, as it relates closely to online social media and the desire of individuals to spread information. The desire to spread information and the theorized reasons behind that desire are crucial to both existing models and the development of new information spread models proposed in future chapters.

3.1 Philosophy, Science, and Information Spread

The historical journey from early philosophy to modern social theory and information spread research is summarized (briefly) in Figure 3.1. Here, we focus on four major periods of philosophical thought and scientific development: the ancient world, the medieval world, the early modern world, and the contemporary world.

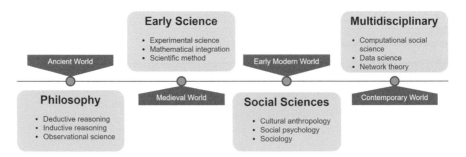

FIGURE 3.1: Historical evolution of information spread research.

3.1.1 The Ancient World

Like many academic topics, the study of information spread has its roots in philosophy. In particular, Platonic and Aristotelian philosophy set the stage for scientific inquiry. According to Plato (and other contemporary Greeks), humans were born with the knowledge of everything, and through learning, this knowledge was unlocked. Any knowledge gained via observation and experiments was only seen through the lens of our senses. An abstract, ideal form of everything exists that we simply may not be able to perceive fully. As such, Plato saw empirical knowledge as an opinion and believed pure knowledge can be obtained only through deduction. In deductive reasoning, premises are taken and then reduced to conclusions based on logic. This is at times called the "top-down" approach to reasoning.

In contrast, Aristotle believed knowledge could only be gained by comparing and building upon what was already known and observed. This thought process forms the basis of inductive reasoning. Sometimes called the "bottom-up" approach, it takes premises and draws likely conclusions based on that evidence. Inductive reasoning is still the cornerstone of many of the scientific theories and discoveries throughout history [17]. Aristotle is often credited as having initiated the scientific method, which is simplified in Figure 3.2.

Reason teaches us to discover, question, and innovate. These philosophical methods of reasoning created the fundamental ways in which scientific disciplines developed. Ultimately, both forms of reasoning have their place in the modern interpretation of the scientific method. Scientists use inductive reasoning it to develop hypotheses and theories. After theories are established, deductive reasoning allows for the application of those theories to context-specific situations [18].

3.1.2 The Medieval World

The "western world" was mostly stagnant after the fall of the Roman Empire with some exceptions. However, due to the earlier conquests of Alexander the Great (Aristotle's pupil), many Arabic-speaking cultures absorbed and

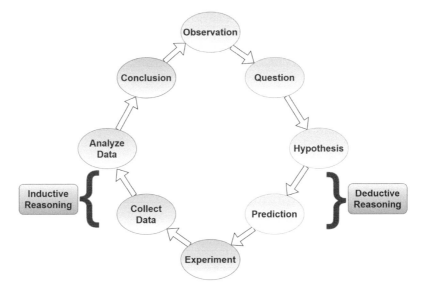

FIGURE 3.2: Overview of the scientific method.

assimilated Greek heritage, including natural philosophy, mathematics, and engineering. Additionally, the writings translated from Greek thought provided the foundations of scientific inquiry. The Golden Age of Arabic science (coined due to most works being translated into Arabic) saw the development of early algebra, algorithms, the widespread use of pens and printing, and explanations of how disease spreads [19]. These discoveries paved the way toward Europe's Renaissance and Enlightenment eras and saw further refining of the scientific method and several collaborative efforts between scientific and religious thought to understand the nature of reality better.

3.1.3 The Early Modern World

Later, the philosopher Francis Bacon helped connect Aristotelian and Platonic thought, reasoning that the universe was too complex to explain using Aristotle's methods alone. Platonic idea was required for great leaps in knowledge to occur. He also significantly advanced experimental science through the belief that real-world observations can be tested through research to determine their validity. Still, that inductive reasoning supplemented this research to generalize the findings to the aggregate population.

Galileo, mostly famous for his carefully designed astronomy observations and experiments, knew that mathematics and physics were essential in describing several natural phenomena. He is also credited as the first to have mathematical modeling play a critical role in the scientific method. As our modern view of science began to take shape, the idea of science and theology as separate entities also gained general acceptance during this period [17].

As science eventually began to gain positive appeal and great strides were made in astronomy, physics, chemistry, and biology, there was a call to see equal advancements in the study of society and human behavior through the sciences. In time, the distinct disciplines of economics, political science, cultural anthropology, sociology, social psychology, and social statistics emerged as necessary and widely recognized areas of scientific research and development. No longer were these topics subsets of philosophical thought, but instead as scientific disciplines in their own right [20].

3.1.4 The Contemporary World

Specialization within various fields of the social sciences became increasingly prominent, just as with contemporary physical and biological sciences. Both contrary and parallel to this trend, however, was the recognition of inter-disciplinary approaches that utilize knowledge and methods from different disciplines. These social science fields, all dealing with human interaction and communication, began to benefit from statistical, mathematical, and modeling techniques, adding new quantitative methods to supplement research and apply it to modern problems.

More recently, multidisciplinary approaches have seen the use of network theory and modern social theory to study and analyze not only how people are connected, but also why they are connected on a social and psychological level. In contrast to an interdisciplinary approach, multidisciplinary approaches involve several people working together, each drawing from their discipline of specialty. Multidisciplinary research areas, such as computational social science, blend social theory with mathematical modeling, and quantitative computational analysis to describe social interactions and communication in new and exciting ways.

The advent of modern computing devices (both traditional and mobile) and communication technology has brought the internet to commonplace usage. With these tools and methods, interest in information and its spread can be researched and taken advantage of like never before. Social media sites allow rapid communication. Data can be collected from any number of sources on network activity, whether it is online shopping or social media post sentiment. With a data-driven internet, mathematical and machine-learning models can potentially predict how information is spread throughout a social group with reasonable accuracy. Time will tell what all of these contemporary advancements will progress toward, but none of them could have happened without the philosophical and scientific developments that preceded them.

3.2   Social Theory and Social Networks

Early concepts of social theory and interactions have been studied decades before anything even resembling social media as we know it today by researchers. Over time, researchers began to summarize and consolidate prior works in psychology and sociology [21]. This led to the presentation of a social exchange theory to examine the exchanges between individuals and small groups, essentially stating that each individual behaves in accordance with a cost-benefit assessment of potential social interactions [22]. The overarching concepts, while not intended for modern online social media interactions, perhaps still hold true.

Other pioneers in information spread theory extended the concepts present in social exchange theory to online social media interactions. By comparing those who generate online information to those who consume, share, and comment on it, it was noted that online users are typically watchers and not producers, but still act under a risk versus reward mentality of sharing, producing, or commenting on content [23]. Intuitively, this makes sense. Out of all of the websites, video sharing sites, blogs, and discussion forums we visit on a daily basis, how much original content is each of us personally contributing versus only viewing or redistributing?

With the evolution of information spread medium, the theories that attempt to describe and examine person-to-person communication must also evolve. Some contend that the "media is the message". That is, the medium by which information is communicated within society has a much greater influence on a group than the specific content of the message being communicated [24]. With a decrease in the importance of traditional newspapers, television, and word-of-mouth spread, internet news, blogs, and social media have emerged as a vital vehicle for news spread. At first glance, it is easy to agree with this line of thinking. The advent of the internet has changed our lives and how we get information, yet it has had little effect on the information itself in most cases.

How information propagates has become an increasingly important topic as widely accessible digital news and media overtakes traditional news sources. This propagation, especially between different individuals or groups, is often referred to as "information diffusion". Many researchers are focusing on the topological aspects of networks to study these phenomena. For example, several researchers examined online information diffusion's role in sharing information with friends using a large-scale field experiment. By examining the role of both strong and weak ties within a large network, they determined that while strongly tied individuals are more influential within a group, weakly tied individuals are responsible for information diffusion between groups or networks [25].

Other research focuses on the content of the information message as a significant factor in its ability to spread. In one instance, an algorithm was developed and evaluated through the use of Twitter hashtag extraction to demonstrate that the content itself has a significant impact on its ability to spread [26].

Information spread in online social networks can be viewed as having four main components: the actors that see and spread the message, the content of the message itself, the network topology through which the message must spread, and finally, the processes by which the message diffuses throughout the network topology. Very diverse topics can predict the popularity of a message within a community. In contrast, low diversity of messages tends to increase the influence of a single individual spreader. Finally, the network community's viewpoint influences information through social reinforcement and homophily to "trap" information in and out of community nodes. As a result, early-stage information does not diffuse as an infectious disease as most models assume [27].

Using an empirical study of news spread on social media networks such as Digg and Twitter, researchers extracted data from the social network sites to demonstrate the critical role they play in information spread and how the network structure affects the information flow [28]. No explicit dynamics or modeling is given, but they demonstrate the importance and value of the empirical analysis of social media sites to support qualitative social theory with respect to online communities.

In mathematical modeling, epidemiology-based approaches tracing the spread of rumors within a population have been slowly evolving through the works of Daley and Kendall (DK model) and later by Maki and Thompson (MK model). Even marketing models have found a home in the study of information spread. After all, what is marketing if not spreading a specific message about a product, service, or event in the hopes that others do the same. Mathematical models such as these and more will be discussed in detail in later chapters.

3.3   Social Exchange Theory

Fundamental to the existence of online social media is the requirement that individuals or groups create and communicate content; otherwise, there would be no information to spread [23]. Social exchange theory emerged from sociology studies, which sought to examine the exchange relationships between individuals and small groups [22]. In social exchange theory, individuals act in accordance with a subconscious cost-benefit analysis type mentality subjective to the individual. If a social behavior is deemed too costly, such as insulting another community member, it will not be acted upon unless there is a greater perceived benefit (perhaps in this case, a benefit of asserting social dominance). Cost-benefit mentality influences our ability to communicate, form bonds with community members, and spread information within the community [22]. Social exchange theory can be summarized as follows:

Social behavior is an exchange of goods, material goods but also non-material ones, such as the symbols of approval or prestige. Persons that give much to others try to get much from them, and persons that get much from others are under pressure to give much to them. This process of influence tends to work out at equilibrium to a balance in the exchanges. For a person in an exchange, what he gives may be a cost to him, just as what he gets may be a reward, and his behavior changes less as the difference of the two, profit, tends to a maximum [21].

Clearly, in a social media driven community, individuals expect to give and gain reputation and influence in the abstract sense via posts, comments, shares, and other popular mechanisms. Expected rewards from these social exchanges may not be monetary, but they certainly can be in the form of sponsored advertisements on social media.

It is also noteworthy that in an online social media environment such as YouTube, far more individuals are consuming content over those creating it. According to the Global Web Index 2009 study on online social media habits in the United States [29], social media users can be categorized as belonging to four main groups: watchers, sharers, commenters, and producers. Watchers (79.8%) view and follow online content only, with no reciprocation. Sharers (61.2%) share, upload, or otherwise spread the content of others. Commenters (36.2%) are individuals who will rate, review, and comment on things like products as a form of contribution without actual material generation. Finally, producers (24.2%) create their own content for any number of reasons, ranging from expression to social recognition [23]. The validity of the groupings requires additional research, but it certainly provides some insight into online social media behavior. That said, promising research has been done to demonstrate affinity, belonging, interactivity, and innovativeness are all included in users' base expectations when utilizing a social media network [30].

3.4   Exercises

1. Explain the differences between deductive and inductive reasoning. Give an example of each.

2. Explain the difference between interdisciplinary and multidisciplinary approaches.

3. Choose a formalized social science. Identify and briefly explain how some external disciplines can be used to redefine problems outside of their normal boundaries or to reach new solutions given complex situations.

4. Explain the connection between pure sociological theory and social network theory.

5. Summarize Social Exchange Theory. Does it seem to make sense in the context of your everyday social media network experiences? Give a brief explanation to support your answer.

4

Social Network Relationships and Structures

> *Frequently consider the connection of all things in the universe.*
>
> Marcus Aurelius

In this chapter, many of the fundamental network relationships and structures are presented in the context of information spread through social media linked individuals and groups. Core social networks relationship terminology, such as symmetry, directionality, intermediary relationships, and network complexity are addressed. Relationship structures, such as dyadic, triadic, and balanced relationships are also summarized in simple terms. The concept of homophily and filter bubbles as communication echo chambers are described using a modern example: the dwindling trust in mass media. The chapter concludes with an overview of several popular social network visualization and analysis software tools. These tools are useful for exploring and understanding the relationships and structures presented in this chapter as well as in social network analysis.

4.1 Social Network Relationship Overview

In social network analysis, each community member is treated as a node, and their communication with other members is treated as a link or connection between nodes. Social networks are analyzed at varying scales, but the primary purpose of social network analysis is to ultimately utilize mathematical models to study the structure, development, and evolution of the social network [31]. Fundamental to social network analysis is its structure, as that will dictate the efficiency by which information can be spread throughout a social network group. The advantage of a social network analysis method of examining social media groups is its ability to quantify relationships mathematically. Social network analysis preliminaries are discussed further in the next chapter.

4.2   Core Social Network Relationships

Critical to the discussion of modern information spread is the concept and influence of social networks. First, let us define a general network. A network is a set of objects or nodes along with a mapping or description of the relationship between the nodes [32]. A social network, then, is a set of individuals related in some way so that their relationship can be mapped or traced.

4.2.1   Symmetry

Consider the most basic case of two friends: a symmetric relationship, in which two friends are linked to each other as shown in Figure 4.1. In this case, *individuals* 1 and 2 are linked within a social network. Specifically, they are symmetrically linked because each has two-way mutual communication with the other. In traditional human interactions, this is the most common social network mapping of a face-to-face communication between friends. Formally, this mapping of individuals is called a "sociogram".

FIGURE 4.1: Individuals 1 and 2 in a simple symmetric relationship.

4.2.2   Directionality

In Figure 4.2, a third individual is added to the network. Notice that the new actor, *individual*-3 has a symmetric relationship with *individual*-1, but is only singularly directional to *individual*-2. This directional-sensitivity in a network relationship is known as directionality. Perhaps *individual*-3 is a writer. While *individual*-2 is being influenced by the information spread from *individual*-3 as he reads his work, *individual*-3 has no direct knowledge of or contact with *individual*-2, while *individual*-1 is acquainted with and talks to both of the remaining people.

4.2.3   Intermediary Relationships

What happens if a fourth individual who is only acquainted with *individual*-2 enters the network map as shown in Figure 4.3? While *individual*-4 and *individual*-3 are symmetrically linked, she has no direct relationship to others

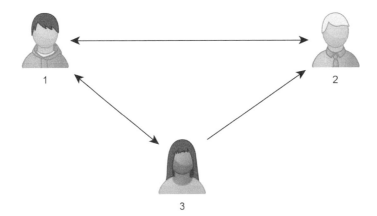

FIGURE 4.2: A three-node relationship mapping of individuals 1, 2, and 3.

in the network. She is said to have an "intermediary relationship" to the rest of the network. In this case, the intermediary is *individual*-2 who serves as her link to the rest of the social network.

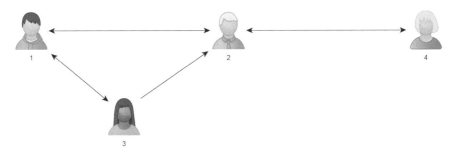

FIGURE 4.3: Individual 2 acts as an intermediary.

4.2.4 Complex Networks

As more people are added to a social network, their interrelationships become increasingly complex. Notice the variety of symmetric and unidirectional relationships in Figure 4.4. By only adding a few more people to our social network, each with their relationship links, the social network has grown complex enough that it proves difficult to trace and predict how information might spread between individuals on opposite sides of the network map.

In the next sections, we will address social networks in more detail about their structure and formalized descriptive elements.

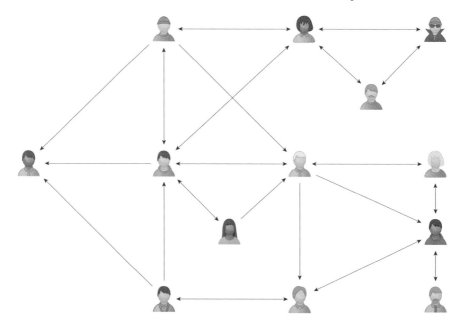

FIGURE 4.4: A simplified complex social network.

4.3 Homophily and Filter Bubbles

Meaning "love of the same", homophily is a term coined in the context of social theory by Lazarsfeld and Merton in 1954 [33]. Mainly, it expresses the concept that similar individuals (or groups of individuals) tend to be drawn together within networks, with closer similarities resulting in closer network bonds. Likewise, people or groups with dissimilarity will tend toward involvement in completely separate social networks. Additionally, these network groups often induce positive feedback into themselves due to similarities between members, making the group "likeness" bonds increasingly stronger [32].

In a modern popular social media context, this phenomenon is often known as a filter "bubble" [34]. Similar to the concept of a social "echo chamber", a social filter bubble is the isolation of individual thoughts, perceptions, and news from opposing viewpoints due to their current belief systems, social media circles, and internet search tendencies. Evolving non-transparent technology has made filter bubbles increasingly intense, as personalized news streams, ads, and searches begin to dominate typical internet activity. Growing concern has arisen as to whether this trend is harming democratic ideals as these concepts enter public consciousness [35] following the recent, social media internet attributed, 2016 United States presidential election results. Additionally, the knowledge of the existence of fake news and filter bubbles has eroded some public trust in traditional television, newspaper, and internet journalism. The graph in Figure 4.5 exemplifies an increasing trend of widespread distrust of mass media over the years in the United States.

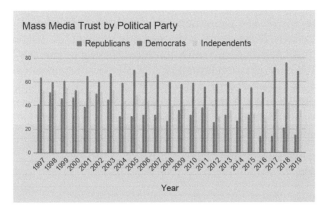

FIGURE 4.5: U.S. trust in mass media trends. Courtesy of [36].

4.4 Dyadic Relationships and Reciprocity

In the realm of sociology, the most straightforward grouping consists of two individuals, also known as a dyad. An example of this might be a teacher and a student. Both have a connection to one another, as they interact with and influence others within a small two-person network. In the case of the teacher-student example, there exists "reciprocity", between the two individuals for the reasons mentioned above. In personal interactions, dyadic reciprocity is frequent. Still, once online social media interactions enter the picture, it is not hard to imagine several typical situations in which non-reciprocal relationships dominate. For example, consider an internet blogger with several hundred followers. The blogger may follow and reply to some of her readers, but for the most part, the blogger is not interacting with the readers, and the relationship is purely one-sided. Returning to the teacher-student dyadic relationship, if the teacher lectures material and the student does not actively participate in course discussion (if any), then there exists no reciprocity in the relationship. In networking terms, these relationships can be called "directed", as there is a one-sided, non-mutual connection between the individuals.

4.5   Triads and Balanced Relationships

Let us expand the simple person-to-person relationship to three individuals, each existing within the same network. With the addition of a third person, network analysis can truly begin because a society (however small) has emerged [37]. A "triad" forms with the introduction of the third individual, and the complexity of the relationships is considerably increased. Consider three individuals: person A, B, and C. Person A is a good friend with person B, reciprocally. Person C is friend with person B but does not know person A; however, person C follows the blog posts of person A due to common interest but is not reciprocally followed.

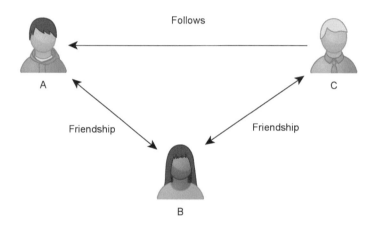

FIGURE 4.6: Example of a simple triad relationship.

As the network grows in size by even a small amount, it becomes more complicated due to the reciprocity of relationships, the presence or lack of intermediaries, and the number of individuals within the network group. For further understanding, the concept of *balance* can be considered. Balance within a triad network can be formalized as follows: *"In the case of three entities, a balanced state exists if all three relationships are positive in all respects, or if two are negative and one is positive"* [38]. It can further be argued that groups naturally tend toward these balanced triad states. As an example, consider a triad where two individuals dislike the third triad member. It is likely then that the two disliking members will like one another, perhaps due to shared ideologies or opinions that cause them both to dislike the third person of the network. Eventually, the third person may even become isolated from the group or network entirely [32].

TABLE 4.1: Social network analysis and visualization tools.

Software	Features
Cuttlefish	detailed visualizations, interactive manipulation, TeX support
Cytoscape	data integration, analysis, and visualization, plug-in support
Gephi	interactive visualization for complex networks and systems, dynamic graphs, open-source
NetworkX	create, manipulate, and study complex networks, Python package, random network generation
NodeXL	Microsoft Excel template, open-source, explore network graphs, social network specific plug-ins
R	general purpose analytics, extensive libraries for social network analysis
SocNetV	analysis and visualization, user-friendly, network construction, cross-platform, open-source

4.6 Social Network Analysis Software

Several social network visualization and analysis software tools exist to aid in understanding and interpreting data collected from social media systems. Some of the major packages and tools for social network visualization and analysis are shown in Table 4.1. Applications of such software includes usage in sociology, finance, biology, network theory, and more. It can be argued that the usage of modern software is not only useful, but required. The complexity of many networks (sometimes with several thousands of nodes or more) makes unassisted visualization and analysis impossible.

Consider Figure 4.7, a randomly generated network of fifty nodes, each representing an individual in a social network. These network sociograms were created with the widely used Social Network Visualizer software (SocNetV). A snapshot of the tool user interface is found in Figure 4.8.

Note that the nodes in the network vary significantly in relation to one another. Some nodes are sparsely connected, others are very dense and connected to several neighbors. Additionally, clusters of nodes in close connection and proximity are also visually apparent.

Figure 4.9 gives reports on the network degree of centrality and clustering of nodes.

For many network development and analysis software tools, random networks can be created given user-input parameters. Specific data can also be imported for the visualization and analysis of real-world data sets. Readers are encouraged to try one or more of these tools to create random networks or importing data to get a sense of visually identifying clustering, density, degree centrality, polarization, and other metrics.

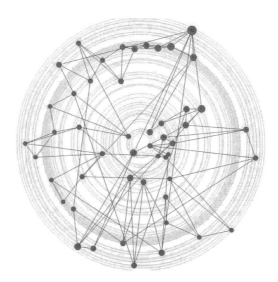

FIGURE 4.7: Randomly generated fifty-node network.

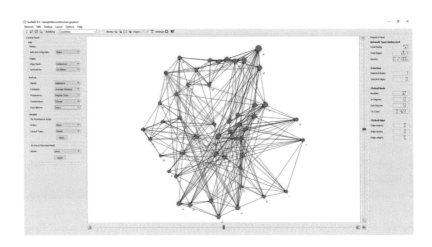

FIGURE 4.8: User interface for the Social Network Visualizer tool.

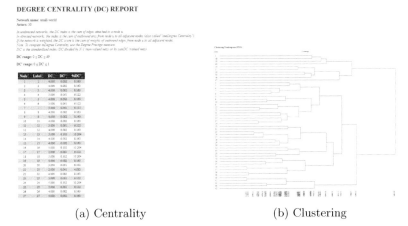

(a) Centrality   (b) Clustering

FIGURE 4.9: Social Network Visualizer: centrality and clustering.

4.7   Exercises

1. Provide an example of a filter bubble in the context of modern-day geopolitics.

2. Give an example of a dyadic relationship. Does reciprocity exist in your example?

3. Consider Figure 4.6 as a sample triad relationship. Discuss the concepts of reciprocity and balance in this context.

4. Filter bubbles are of increasing concern for mass communication and socialization in a digital media age. Give an example of a filter bubble one might encounter and draw a sociogram of your example. Identify the major nodes that influence information spread within your sample network.

5. Choose a software tool from Table 4.1 (or a similar tool of your choosing). Generate a random 100-node directed small world network. Once your network is developed, generate a degree centrality report and determine the adjacency matrix.

5

Social Network Analysis

“ *A hidden connection is stronger than an obvious one.* ”

Heraclitus

This chapter will review some basic social network analysis fundamentals. Understanding the essential theories, concepts, and terminology is critical for further discussion of information modeling and control within the scope of online social networks. The network concepts of density, structural holes, strength of ties, centrality, and distance are briefly explained with visual examples and some mathematical representations. Small world networks and polarization are discussed and are of particular interest when examining online social media groups. Using a simple three-node network group, the relationship between a network configuration and its adjacency matrix is examined. To conclude the chapter, an example of a directional sociogram and its accompanying adjacency matrix for a sports club is given to pave the way toward practical social network analysis applications.

5.1 Density and Structural Holes

Network density is defined as the number of direct actual connections divided by the number of possible direct connections in a network. A "potential connection" is a connection that could potentially exist between any two nodes, although it may not actually be connected. An "actual connection" is one that actually exists [39]. Equation 5.1 gives the mathematical calculation for network density, where n is the number of nodes in the network. Figure 5.1 visually compares a sample sparse and dense network.

$$Potential\ Connections = \frac{n(n-1)}{2}$$

$$Network\ Density = \frac{Potential\ Connections}{Actual\ Connections}$$

(5.1)

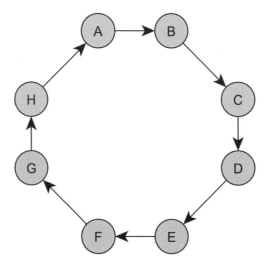

(a) Sparse network.

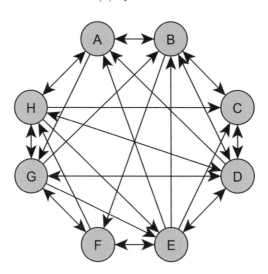

(b) Dense network

FIGURE 5.1: Comparison of sparse and dense networks.

A real-life group such as a class or club would typically be reasonably dense because each individual is usually acquainted with (or directly connected to) their classmates or group members. Similarly, online groups with high levels of direct communication such as family social media groups or online game "guilds" of sufficiently small size will be relatively dense. Higher levels of density often come paired with an increase of information spread and a sense of community along with the resultant inter-group social support structures. By their nature, small networks tend to be denser than large social networks. It's easy to know everyone in a class of twenty individuals, but knowing everyone in an entire school becomes increasingly unfeasible.

In direct contrast to the concept of density is what Burt refers to as "structural holes" [40]. Imagine two dense networks comprised of individuals that mostly know one another, and a single individual is a part of both groups, being their only common connection. If we imagine the networks combined into a single, larger grouping, there exists a structural hole within the new network, centered around the cluster-bridging individual. Figure 5.2 illustrates a single connecting individual bridging two clusters within the same social network. One may naturally wonder why these groups are in the same network and not divided, but several real-world examples of structural holes in social networks are common. Politically different groups within the same country, rival teams within a sports league, and college courses taught by a single professor at two different schools are all good examples of social network structural holes.

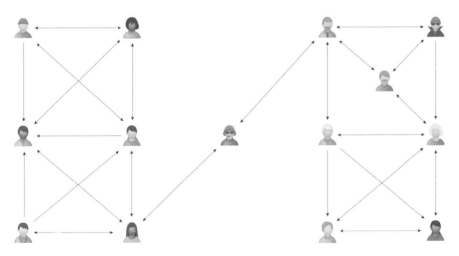

FIGURE 5.2: Illustration of a structural hole.

5.2   Weak and Strong Ties

The concept of weak ties is closely related to that of a structural hole, in that weakly tied social networks are linked by a few bridging individuals, such that two or more distinct group clusters can be readily identified. Practically speaking, weak ties help prevent large networks from being completely fragmented by facilitating the spread of information between segments. However, other factors help define a tie strength, such as the length of time individuals are acquainted, level of interaction, and how close in friendship or acquaintance individuals subjectively feel toward one another [32]. Pairs with strong ties might be friends or family members, while those with weak ties are more likely simple acquaintances, coworkers, or neighbors. Especially in online social networks, weak ties can play a critical role in information diffusion. Strong ties can be seen in the opposite fashion. They facilitate reinforcement of group values and tend to feed the same ideas and culture back into the group. An example of a network with both strong and weak ties is shown in Figure 5.3, where solid lines represent strong ties and dash lines signify weak ties.

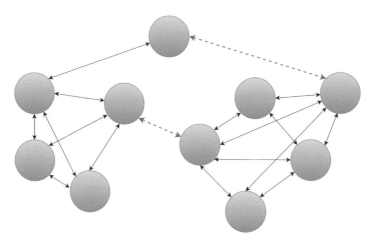

FIGURE 5.3: Strong and weak ties.

5.3   Centrality and Distance

In simplest terms, centrality describes how connected a node is to the network [23]. A centralized node will be highly connected to several other important nodes and hence have easier access to a number of network members when compared to a low centrality node. As there are many ways to define the importance of a node based on its connectivity, there are multiple methods used to define centrality quantitatively. Popular centrality measurements including degree centrality, closeness centrality, betweenness centrality, eigenvector centrality, and Katz centrality, to name a few. In Figure 5.4, nodes of high centrality are readily apparently by their high level of connectivity and importance to the network structure. The removal of these nodes would considerably change the network structure, while outlier nodes with few connections would keep the basic structure of the network intact.

FIGURE 5.4: A random network demonstrating centrality and distance.

Degree centrality can be thought of as a node's risk of catching whatever information (in this context) is flowing through the network immediately and represented mathematically as follows:

$$C_D(v) = deg(v), \tag{5.2}$$

where v is the node of interest. Additionally, degree centrality can be expanded to the entire network group to measure network centrality, or the degree to which the network is centralized is determined by:

$$C_D(N) = \frac{\sum_{j=1}^{|V|}(C_D(v^*) - C_D(v_i))}{H} \tag{5.3}$$

where v^* is the highest degree node of network graph G. H is defined as:

$$H = \sum_{j=1}^{|Y|}(C_D(y^*) - C_D(y_i)),\qquad(5.4)$$

with y^* as the node with the highest degree centrality in the network Y that maximizes H. Mathematically, closeness centrality (the most intuitive measure) is calculated as:

$$C(x) = \frac{N-1}{\sum_y d(y,x)},\qquad(5.5)$$

which is the reciprocal of "farness", where N is the total number of nodes in the network and $d(y,x)$ is the distance between the x and y vertices [41].

Related to centrality is the idea of "distance" between nodes of a network. Also known as a geodesic distance, network structure distance is formally defined as the distance between any two nodes is the length of the shortest path via the edges or binary connections between nodes [42]. Typically, distance is calculated using breadth-first traversal [43] or Dijkstra's algorithm [44]. Consider Figure 5.4 again and note that the highly connected central nodes can reach nearly any other node in only a few steps, by following their connection lines. Direct neighbors will be reachable in only a single step, while individual outer nodes will take two of three steps. In contrast, less centralized nodes will need, at minimum, several steps (perhaps passing through these centralized nodes) to reach other outlying network members. The centrality and distance concepts become increasingly important when discussing large populations such as cities or political groups. In the context of online social media, these principles remain true. An entrepreneurial or political leader will be a centralized individual within a country when making an online post or announcement, just as they would be if their information were to spread in traditional media outlets such as television and newspapers. On social networking sites such as Facebook, "friends of friends" are a greater network distance from an individual than their core friend group, so receiving and relating information becomes slower and more difficult.

5.4   Small World Networks

Consider a network in which there is no overlap between each individual's personal networks if taken as a series of simple nodes and their direct neighbors. In this scenario, each new individual added to a network brings in an entire group of new network members that they alone have acquaintance with. It's easy to see that networks organized in this fashion can attain extensive reach by adding only a few members. However, such groupings are not common in typical real-world networks, particularly when discussing online social networks. Friends have common friends (or friends of friends) that all know each other from the same college or club. Coworkers know many others in the office and do not befriend each other in entirely isolated groups. There are usually several non-unique individuals with relationships from overlapping sources (via a combination of strong and weak ties). These types of networks are known as "small world" networks [45].

Small world networks are perhaps the most commonly discussed and analyzed due to both their limited scope and realistic inter-connectivity. Such networks can be imagined as several sets of highly connected teams with some connectivity between members of other teams. A sample small-world network is shown in Figure 5.5.

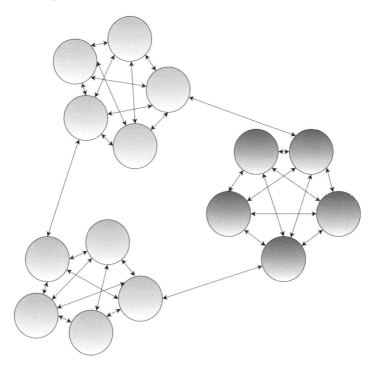

FIGURE 5.5: Sample small world network group.

There are several advantages to small-world networks. In fact, they are generally regarded as highly robust. For example, any one random node will have a reasonably short path to another node. If something happens to one subgroup or it is somehow cut-off from the others, it is not entirely isolated (though it will potentially require additional steps to reach). Due to the high connectedness of the smaller groups, they can communicate and work quickly and efficiently, while still having a connection with other groups. It's easy to see how a small-world network configuration would be beneficial in a business or production environment. In biological groups, it helps reduce potential damage a virus or genetic mutation might have on a population, as infected subgroups are less connected to other groups. Examples of small-world networks include power grids, social network influencers, and voting groups within a political party.

With all of the advantages of small-world networks, there are also some key potential disadvantages, especially in relation to social media and information spread. Small network groups tend to resist change since members of their subgroup have a strong influence over each member (regardless of what true or false information might be coming from other subgroups). This can allow misinformation to be widely believed within a smaller and closer group and makes it very difficult to change each member's beliefs. Social media "echo chambers" based on culture, socio-economic class, and political ideologies are allowed to thrive in such a network environment. This is why many social media users will continuously encounter the same talking points and further reinforce existing beliefs. How often have we asked ourselves why our friend groups and colleagues seem so reasonable, but others are posting nonsense?

5.5   Clusters, Cohesion, and Polarization

The idea of social network clusters are closely linked to Charles Cooley's concept of primary groups:

> *By primary groups, I mean those characterized by intimate face-to-face association and cooperation. They are primary in several senses, but chiefly in that, they are fundamental in forming the social nature and ideals of the individual. The result of intimate association, psychologically, is a certain fusion of individualities in a common whole, so that one's very self, for many purposes at least, is the common life and purpose of the group. Perhaps the simplest way of describing this wholeness is by saying that it is a "we"; it involves the sort of sympathy and mutual identification for which "we" is the natural expression. One lives in the feeling fo the whole and finds the chief aims of his will in that feeling [46].*

In many ways, clusters are similar to Cooley's primary groups, but they do not overlap. Under cluster categorizations, one cannot be a member of multiple clusters at once. Sometimes there exist hierarchies and organization by which members identify themselves, but oftentimes (especially in vast social networks), formalized categorizations can get messy and blurred even if they technically exist [32]. Based on datasets created by Lada Adamic in 2004, Figure 5.6 shows two distinct politically oriented blogs: liberal and conservative, forming two distinct (highly polarized) clusters within the online social bloggers' network [47].

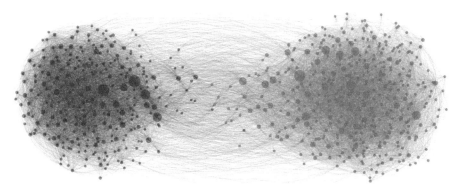

FIGURE 5.6: Polarization of the U.S. Political Blogosphere. Courtesy of [48].

Cohesion is a measure of network group connectivity in social groups. It defines the minimum number of individuals that must be removed from the group to cause it to dissociate. Ideally, a highly cohesive group will be connected to several members within the same cluster in a network such that severing individuals from the group does not cause the cluster to break apart to any substantial degree. Cohesive primary groups within a larger social network are often casually referred to as "cliques". The strength or cohesion of cliques can be measured by their ability to pull together as a group to resist disruptive forces directed toward the clustered network group [49]. For example, if someone challenges the beliefs and norms of a cohesive cluster, it will join together to reinforce those beliefs and norms.

In modern social network commentary, cluster polarization is a hot topic. Figure 5.6 exemplifies a highly polarized political community in which the vast majority of online social network members blog with strong ideological tendencies, usually in direct opposition to another strongly cohesive cluster. Most members are either firmly liberal or conservative, with only a small section of the network acting as moderate bloggers. The concepts of homophily and filter bubbles discussed earlier come into play in scenarios where a network is polarized, as members surround themselves with information with which they already agree. Concerns have been expressed over the dangers of this trend, especially with the advent of online social media sites where members of a cohesive clique can easily fall into their own bubbles of personalized news feeds, search recommendations, and YouTube video programs [50]. Modeling and attempts at controlling highly polarized groups will be addressed in later parts in detail.

5.6   The Adjacency Matrix

In the previous sections, some network relationships were examined on a high level, but there was no mention of how to represent them mathematically. One way to describe a network and its interrelationships is the adjacency matrix.

Again, let us consider the three-node sociogram from the previous chapter, as shown in Figure 5.7. Notice that each node pair of individual relationships has two elements of interest: direction and the presence of a connection. An adjacency matrix can be formed from the simple social network structure to show the mathematical relationships between each pair of the networked group. In the sample adjacency matrix presented in Table 5.1, 0 represents no connection between the paired groups and 1 represents a connection. It should be noted that the connection is directional. While one person may be connected to an adjacent individual, that second individual may not have a connection to the initial person. Unidirectional connection situations are common in social media, in which one user might follow a celebrity or social media "influencer", but not be reciprocally followed.

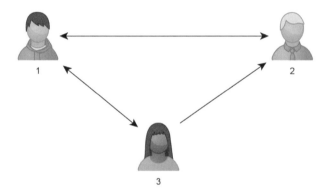

FIGURE 5.7: Revisiting the three-node relationship map.

TABLE 5.1: Adjacency matrix of a sample 3-node network.

	1	2	3
1	0	1	1
2	1	0	0
3	1	1	0

The tabular matrix can be rewritten as a standard matrix for future mathematical manipulation.

5.6.1   Example: A Fencing Club Sociogram

A typical fencing club is broken up into three subgroups based on the club member's primary weapon: foil, epee, and saber. Naturally, members of each weapon subgroup know each other. During the course of club practice days, friendships are developed between members that are independent of weapon type based on common interests, personality type, etc. Additionally, there is a central figure: the fencing coach. The coach knows all of the members and interacts with them regularly during practice times. An overview of the fencing club's connectivity is shown through the sociogram in Figure 5.8.

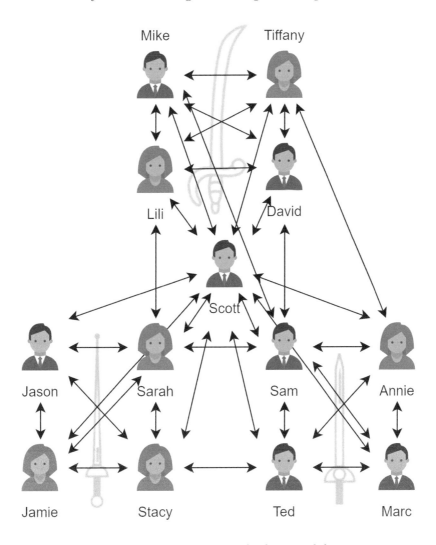

FIGURE 5.8: Sociogram of a fencing club.

Notice that all of the relationships are bidirectional. This makes sense in the example's context, as the in-person nature of the interactions would imply that if one person is acquainted with another, the other person would know them back. Also, notice that Scott, the fencing coach, is the central figure connecting the three subgroups. Though not shown, there can also be strong and weak ties associated between individuals for added complexity. The adjacency matrix can easily be determined by visually tracing the directional connections between members, as shown in Table 5.2.

TABLE 5.2: Adjacency matrix of a fencing club.

	Annie	David	Jamie	Jason	Lili	Marc	Mike	Sam	Sarah	Scott	Stacy	Ted
Annie	0	0	0	0	0	1	0	1	0	1	0	0
David	0	0	0	0	1	0	1	1	0	1	0	0
Jamie	0	0	0	1	0	0	0	0	1	1	1	0
Jason	0	0	1	0	0	0	0	0	1	1	1	0
Lili	0	1	0	0	0	0	1	0	1	1	0	0
Marc	1	0	0	0	0	0	0	1	0	1	0	1
Mike	0	1	0	0	1	0	0	1	0	1	0	0
Sam	1	1	0	0	0	0	1	0	1	1	0	1
Sarah	0	0	1	1	1	0	0	1	0	1	1	0
Scott	1	1	1	1	1	1	1	1	1	0	1	1
Stacy	0	0	1	1	0	0	0	0	1	1	0	0
Ted	1	0	0	0	0	1	0	1	0	1	1	0
Tiffany	1	1	0	0	1	0	1	0	0	1	0	0

5.7 Exercises

1. Discuss if a sparse network or a dense network will be more suitable for information spread? For misinformation spread?

2. Give an example of a real-life social media network having both strong and weak ties.

3. Create a random map using Social Network Visualizer (SocNetV) or similar software having 50 nodes. Compute the centrality and distance of this network.

4. Draw a sociogram of two or more intersecting groups in your daily life (friends, clubs, family, etc.). Explain your sociogram and how the groups are connected through one or more nodes. Create an adjacency matrix based on your sociogram. Who are the most and least connected individuals?

Part II

Macroscopic Modeling and Information Spread

6

Modeling Basics

> *If you want to know your past — look into your present conditions. If you want to know your future — look into your present actions.*

> Chinese Proverb

This chapter begins with a simple definition of a model along with why models are so important when trying to mathematically and conceptually explore a natural or man-made system. Special attention is paid to the role models play in decision making in industry and advertising environments and the importance of recognizing model assumptions and approximations. Some of the major standard types of models are considered and explained, with a deeper focus on mathematical systems modeling. The idea of microscopic and macroscopic level models are presented and compared. With a basic understanding of models, a set of basic steps used in model development is given with particular highlights given to model validation, which will see more detailed examination in future chapters. Finally, examples of modeling are given in the context of a spring-mass system, a predator-prey system, an RLC circuit, an epidemic model, and a vehicular traffic model.

6.1   What is a Model?

Simply put, a model is a representation of "reality." It serves as a stand-in for an actual system, physical entity, process, natural phenomenon, and more. Why do we use models? Because it is either massively impractical or impossible to study or understand the physical entities or systems directly. While the simple physical act of a boulder rolling down a hill can be directly set up, observed, and analyzed. More complicated examples such as a planet orbiting a star are impossible to manipulate and analyze directly to further our understanding of said planetary orbit; thus, models are needed.

This said, it is crucial to keep in mind that while models are a representation of reality, *models are not reality*. Models do an excellent job of allowing us to observe, discuss, and analyze something indirectly. Models cannot, however, be assumed to behave identically to the actual object, system, or phenomenon that is being modeled. Fringe cases might exist that make the model break down (much like traditional physics fails to model many physical objects and processes at high speeds accurately and for incredibly small physical systems like subatomic structures). Additionally, given the generality of many models, small realities not accounted for in the model or specific to the system or object being modeled might lead to false assumed outcomes if the model were to be followed blindly. Still, models are highly valuable tools to help both conceptually and mathematically understand and manipulate the world around us.

6.2   Models in Decision Making

Typically, models are used in industry, engineering, economics, advertising, and more in order to predict system response for specific input. Given a good model, one can input parameters specific to an object or system and receive an expected outcome. For example, take our example of the stone rolling down a hill. Assuming we had a reliable model for how a boulder rolls down a hill (we do!), we can input as parameters the boulder's size and weight, the shape and size of the hill, and the roughness of the terrain, to have a pretty good idea where and when the boulder will be at a given time. Will it be exact? Probably not, but it will provide a prediction of the most likely future reality.

If a future reality can be predicted within reason, then it makes sense to use models to make future decisions. In fact, models are heavily used for decision-making. Major modern retailers such as Amazon have a pretty good idea of the resources (be it inventory, personnel, or capital) that must be invested into their various system segments to have the best efficiency possible. Even though they don't *know* what anyone might order on any given day, it can be predicted using models to optimize time and resources. For example, more drivers and warehouse workers are needed around gift-giving holidays. While this seems obvious, less obvious is predicting how many more drivers and workers are required without hiring too many. The history of previous sales at holidays are the indicators that will eventually be built into prediction models to influence hiring decision making.

Let's return to the simple boulder rolling down a hill example. If one plans to build their house nearby, it would be wise to check how the boulder might fall and ultimately land if something (like an earthquake) caused it to begin moving. Obviously, there are no guarantees in a real system that is simply being modeled, but it would definitely be a good idea not to build the house anywhere the boulder is predicted to cross or land. Although this example may seem trivial, many houses are built in areas with reasonably predictable problems like landslides, flooding, or fires that can be modeled to inform builders and buyers of the future risks they are undertaking.

6.3 Standard Models

Below are some of the major model types to consider. This list is hardly exhaustive, and new subclasses of models arise as disciplines that utilize models mature.

1. **Science** models are used to model nature. It is a general category that can overlap with others. Climate models, migration pattern models, and astrophysical models are good examples of science models.

2. **Engineering** models are used to model man-made systems. Like science models, this category of models can overlap with other groups. Engineering models are used for bridge construction, electronic circuit design, engine analysis, aeronautical design, safe building demolition, and more.

3. **Physical** models are physical representations of an actual entity. Often this is done because it is either impractical or impossible to examine some things directly. Globes that model the earth or atomic structure models are excellent examples. Physical models are widely used in planning and construction, where having a miniaturized version of a physical entity is useful.

4. **Analog** models use an "analog" system to model the original system or object of interest. For example, circuits can be used to model a machine's input and output characteristics because they behave similarly in many ways. Keeping track of what assumptions are made when using this type of model is key. That said, much of analog model usage has fallen out of favor recently and been replaced with computer software simulations.

5. **Schematic** models include organization charts, process flow diagrams, and block diagrams. They are used in areas as varied as company structure overviews to computer program flowcharts. Circuit and mechanical schematics also fall into this category.

6. **Mathematical** models are used to relate reality to mathematical relationships. They typically take the form of equations. Famous physical relationships such as $V = IR$ and $F = ma$ are basic examples of mathematical models for electronics and dynamics, respectively, though many mathematical models can quickly become complex.

7. **Data-based** (or learning based) models are developed from mass data collection to draw important insights. Although they are non-mathematical

and produce no closed-form expressions, they can complement mathematical models by relating them to real-world data. Machine learning can also be heavily used in the development of these models. Netflix video recommendations, Amazon product suggestions, and targeted advertisements all make use of data-based modeling.

TABLE 6.1: Standard model types

Model Types	Usage	Examples
Science	nature and natural phenomena	migration patterns and planetary orbits
Engineering	man-made systems	miniature structures and vehicles
Physical	physical objects or structures	globes, atomic structures
Analog	similarly behaving system	a circuit to model a machine
Schematic	organizational or functional interconnected symbols	block diagrams, circuit schematics
Mathematical	equations and mathematical expressions	$V = IR$, $F = ma$
Data-based	data collection and analysis	Netflix and Amazon recommendations, targeted ads

It is essential to observe that some of the model categorizations can overlap, particularly with high-level classifications such as science and engineering models. The general model types, their usage domain, and examples of usage are summarized in Table 6.1.

To better understand, suppose we want to model a rubber ball. Indeed, one can be built and examined directly. Suppose we want a giant rubber ball, though, and constructing it would be costly. Additionally, consider the value in having a system in place to analyze an arbitrarily sized ball. In this case, we can use a standard-sized ball as a physical model. If we were interested in conveying how the ball is made or how it looks internally, a schematic model would be appropriate. As a bonus, this type of model can be easily emailed, copied, or distributed in any number of ways. Finally, suppose our interest in the object goes beyond its basic form, and we wish to integrate it into other models or apply physics principles. Here, a mathematical representation of the object or its physical properties would be necessary.

6.4 Models, Assumptions, and Approximations

Earlier, it was emphasized that one should remember that models are not reality, merely a representation of reality. So why are models not always perfect and able to be utilized as if they are the "real thing"? Simple: several assumptions go into the creation and implementation of models. In the boulder model, we assume the boulder is relatively smooth and spherical, not jagged, and oddly shaped. We assume the ground is not frozen or covered in excessive debris. External factors such as weather or wind might be ignored. In traditional Newtonian physics (which can be seen as a system of models), it is assumed that interacting objects are not moving near the speed of light, near a black hole, or subatomic.

Assumptions are an inevitable part of model creation and usage. Otherwise, every model would get bogged down with conceptual and mathematical complexity and lose its usefulness. Knowledge of what assumptions are used in models allows model-users to choose the best models for their needs or adjust them to accommodate for specific uses of the model. It should be noted that while assumptions will be present in nearly every model; it is crucial to make the *right* assumptions. Flawed assumptions will ultimately lead to a flawed model.

Assumptions can be subdivided into general categories:

1. **Operational** assumptions are concerned with how something operates or will be operated.

2. **Environmental** assumptions detail where the system or object is being used. This can be a physical environment or conceptual environment as in the case of sociological and behavior models.

3. **People** assumptions consider what users, bystanders, or operators might be involved in the model and their role in the system.

In addition to assumptions, approximations are also necessary elements of model development and utilization. Consider the force of gravity with which a mass is attracted to the earth, given by:

$$F = mg, \tag{6.1}$$

where F is the gravitational force, m is the mass of the object, and g is the acceleration of gravity. This equation should be very well known to most readers and seemingly trivial as an example, as it is typically *approximated* as

$9.8\frac{m}{s^2}$. However, there are slight variations in the value of g. Not only is the earth not an actual sphere, but it also has varying density based on subsurface geological structures. As such, for any given area, this value is based on a combination of the proximity to the earth's poles and the distance above sea level. Even still, it would be an approximation. As with any model of a complex system, an exact value is impractical or impossible. Still, it is useful to make reasonable approximations to handle general cases, so long as one keeps in mind that special cases may develop that call for modifications to the assumptions and approximations.

6.5   Mathematical Systems Modeling

The majority of this book will focus on mathematical models, specifically, models for social media systems. What separates mathematical models from many others is the ability to apply data and parameters to the model as inputs in order to obtain numeric outputs. Additionally, they can easily combine with and complement existing models as needed.

Another benefit of mathematical models is the ability to apply science and engineering methods to them directly. For instance, if a home thermostat is modeled with a mathematical model (one or more equations in this case), then engineering control methods such as feedback loops can be mathematically applied to the model directly to regulate the home's room temperature. It cannot be done directly with a physical model or schematic drawing, for example.

6.6   Microscopic and Macroscopic Models

So far, we have mostly focused on modeling individual entities. However, it is often useful and even required to model groups acting as a whole. In a previous example, a boulder was rolling down a hill. This can be modeled as a single object. What happens if several smaller rocks are rolling down the same hill after a natural disaster strike or similar situation? It would be impractical and not especially useful to model each rock and combine the models for each when we generally want to know where they will fall. In this case, the concern isn't where each rock falls, but if you built your house in a poor location down the hill! Given these concerns, models of systems or entities can be defined on a microscopic or macroscopic level.

Microscopic level models (sometimes called agent-based models) focus on individual entities or a collection of individual entities. A school of fish is

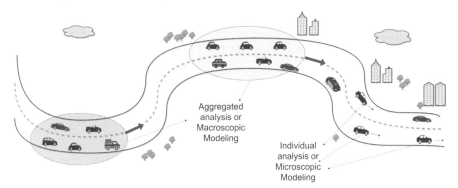

FIGURE 6.1: Microscopic and macroscopic modeling of vehicular traffic.

made up of hundreds of individual fish. Each fish swims at a certain speed with a general movement pattern and responds to the fish surrounding it. This group of fish, as neighbor-dependent individuals, will behave in a certain fashion and can be modeled by using each fish's individual model and accounting for inter-group interactions between the fish. It can be beneficial if we care about any given individual. Perhaps there is a group leader or the physical formation of the fish influences individual behavior. In road traffic, a self-driving car will behave differently from human-driven cars and must be treated differently while modeling. Some users are much more influential on social media services than others due to their number of followers, celebrity status, or expertise in the discussion domain.

Macroscopic level models are concerned with society, the state of society, or a group treated as one entity. By "one entity", it is not meant to imply that the previously mentioned school of fish is modeled the same as one fish, but as an aggregate (or overall combined whole) of all of the individual fish by integrating and accounting for microscopic characteristics. In this case, a macroscopic level model of a school of fish will combine elements of leading fish, boundary fish, center fish, and any other categories to treat the entire school system as a single "average" entity. For mostly homogeneous groups (where all entities are reasonably similar to each other), this can be an advantageous modeling method. It typically simplifies very complex systems into functional groups that can be readily analyzed or mathematically manipulated. Road traffic in this model level is concerned with overall traffic density, speed, and flow, not any individual car. For social media uses, trending tweets, viral video spread, and even how quickly your favorite show gets spoiled online are good reasons to use macroscopic level models. Any individual will make only a small impact on these cases, and it's the population as a whole that is of interest.

Figure 6.1 demonstrates the differences between microscopic and macroscopic models for a sample traffic system.

6.7 Basic Steps to Develop a Mathematical Model

There are many ways to create mathematical models. The method used largely depends on the system being modeled. However, several key steps are common throughout most model development, as follows

1. Identify and quantify objectives

2. Develop and draw a basic conceptual model

3. Validate basic model concepts

4. Perform basic sensitivity and evaluation tests

5. Collect data to validate the model

6. Use the model for various applications

The model development process is summarized in Figure 6.2. The first step of model development is to identify and quantify the model objectives. What is the model trying to do, or what questions is it attempting to answer? How accurate or detailed of a model is required? Who will be using the model, and for what purpose? What will the model's output look like, and what time and space scales will it cover? All of these questions must be answered initially in model creation. It will help focus the direction of model development in order to meet a practical goal.

Next, a conceptual model must be developed. Draw or diagram the model as best as possible and identify what physical laws or principles must be integrated into the model along with the system parameters. Additionally, identify what assumptions or approximations should be used to both simplify and generalize the model for typical use (or for a specific application). The diagram should assist in identifying and writing the basic equations necessary. It will result in a basic model that can be further developed.

Validating the basic model concepts is critical at this point because any oversights will lead to much larger problems in the future. Does the model seem to represent the system being modeled accurately? Does the output seem reasonable in scale, and vaguely match the expected results? If not, return to the conceptual model and make corrections based on these preliminary findings.

After the basic model concepts are verified, preliminary sensitivity and evaluation tests should be performed. For example, a system might perform differently under extreme conditions (hot or cold weather, examining tiny populations, etc.) or behave very differently after a small change of one parameter. Do very high and low values as inputs lead to outputs that make sense in reality, or does the model break down at those ranges? If so, perhaps the model assumptions are flawed, or approximations should be adjusted. Think through and test different typical and extreme values to see where the model begins to fail (ideally using computer software).

To properly validate a model, data must be collected. The sensitivity analysis results should help determine what data and level of detail for that data are needed. Use that data to test the model against reality. Ideally, a very high confidence interval is desired. However, this is much more difficult in systems with a great deal of uncertainly and chaos, such as biological or social systems. The model will never validate perfectly, but it should give clues as to when the model is valid or invalid.

The model can finally be used in practice. During this time, it may need to be re-evaluated or expanded as new cases arise that were unanticipated. That is fine. In fact, it's a necessary step in evolving scientific and engineering paradigms as these fields gain understanding through new research and discoveries.

6.8 Model Validation

Once a model is mostly complete, there are several questions one must ask. How good is the model? Does it compare well with experimental results? Does it accurately match the real system being modeled as closely as practical? Do assumptions or approximations heavily influence the output of the model? These questions all fall under model validation, the processes, and activities intended to verify that models perform as expected and are in line with their design objectives.

There are many ways to validate a model, particularly a mathematical model. This topic will be discussed in detail in later chapters using social media model examples. For now, we will simply say that model validation is the comparison of the model to the actual system, resulting in favorable outputs. If a model reasonably predicts an output given an input, we can say that the model is generally good. For mathematical systems, a simple test of a good model is having low error. If an identical input is sent to both a model and a real-world system, a resulting pair of outputs is produced (one for the model and one for the real-world system). The numeric difference between these outputs is known as the model error. A small error is favorable, while a significant error usually points to a problem with the model, assumptions, or approximations that must be reexamined. Sometimes it may be difficult to validate models in a practical sense, especially with social system models, and special validation schemes must be developed with care.

6.9 Modeling and the State-Space Representation

As system complexity increases, it becomes more challenging to represent multiple system elements in an understandable fashion mathematically. Some

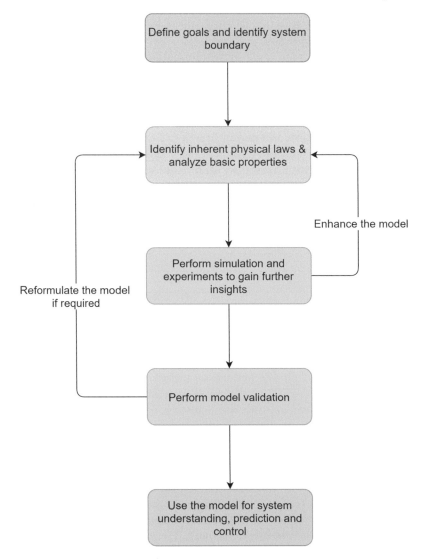

FIGURE 6.2: Summary of model development steps.

equations may be first or second-order differential equations, and the input-output relationship of the system model may not be intuitively clear. It is especially true if one system element influences another and cannot be examined independently. In other words, when system dynamics are coupled or complex, a new way to represent them is useful.

State-space representation attempts to address this issue by representing mathematical models as a set of input, output, and state variables related by first-order differential equations referred to as state equations. Based on the input values and state equations, the state variables update their value at any given time as the system evolves. Similarly, the output variables will update based on the state variable values and output equations. Here, we will take the simplest case, where the dynamical system is a linear time-invariant (LTI) system. In this case, differential equations can be written as simple algebraic equations and matrices. This representation allows for a convenient way to represent and manipulate systems with multiple inputs and outputs.

The general state-space representation of an LTI system has the following form:

$$\dot{\mathbf{x}}(t) = A\mathbf{x}(\mathbf{t}) + B\mathbf{u}(\mathbf{t})$$
$$\mathbf{y}(t) = C\mathbf{x}(\mathbf{t}) + D\mathbf{u}(\mathbf{t}).$$

(6.2)

where:

- $\mathbf{x}$ is the state vector

- $\mathbf{y}$ is the output vector

- $\mathbf{u}$ is the input or control vector

- A is the system matrix

- B is the input matrix

- C is the output matrix

- D is the feedthrough or feedforward matrix

- $\dot{\mathbf{x}}(t)$ is $\frac{d}{dt}\mathbf{x}(t)$

The first equation, $\dot{\mathbf{x}}(t)$ is known as the state equation or "dynamics" of the system. It is one or more functions that describe the evolution of the system over time. The control term of the system, $\mathbf{u}(t)$, is the input to the system, usually in order to influence the system output. This control element can be included in one or more state equations to affect the system beyond its natural evolution. The second equation is referred to as the output equation, which describes the system output. It is important to note that the state variables must be linearly independent. In other words, they cannot be expressed as a linear combination of any other state variables.

For an intuitive understanding of models in state-space form, addition and subtraction of terms in state equations can be interpreted as elements that contribute to a change in the system states. The control term $B\mathbf{u}(t)$, for example, is the addition of a control action to the system in order to change the output. In contrast, the $A\mathbf{x}(t)$ term is the natural evolution of the system, without external influence. Multiplication and division of terms in state equations can be interpreted as an interaction of elements contributing to a change in the system states. Next, we will examine several examples of system models and how to represent these models in the state-space form.

6.10 Example 1: A Spring-Mass System

Suppose we wish to mathematically model a general mass being pulled away from a secured spring. This is known as a "spring-mass" system and is diagrammed in Figure 6.3. First we will identify the elements of the model:

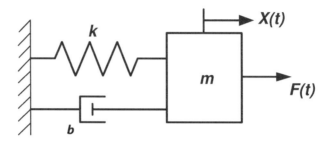

FIGURE 6.3: A spring-mass system.

- m is the general mass of the object
- $x(t)$ is the displacement from equilibrium position
- $F(t)$ is the force being exerted upon the mass
- k is the spring constant that defines its physical properties
- b is the dampener constant (representing friction in this case)
- $\frac{dx}{dt} = \dot{x}(t)$ is the velocity of the mass

These elements can be combined via their interactions with each other. Specifically, the frictional or dampening force is the interaction of the mass's velocity and the properties of the dampening element. The friction will make it harder to move the object. Additionally, the spring is interacting with the displacement of the mass. As the object moves further from its resting point, the spring will tighten and make it more difficult to move the object. From here, we can express these interactions mathematically:

- $b\dot{x}(t)$ is the friction or dampening force
- $kx(t)$ is the effect of the spring on the force moving the object

Now that our model is diagrammed and its key elements and their relationships are identified, we can use existing models and physical laws to proceed. In this case, Newton's second law of motion:

$$M\frac{d^2x}{dt^2} = F(t) - b\frac{dx}{dt} - kx(t). \qquad (6.3)$$

This time-space model can be converted to state-space by replacing the relevant first and second derivative terms with their state-space equivalents and reordering the resulting equation:

$$M\ddot{x}(t) + b\dot{x}(t) + kx(t) = F(t) \tag{6.4}$$

Now assume two state variables $x_1(t) = x(t)$ and $x_2(t) = \dot{x}(t)$, then

$$\begin{cases} \dot{x}_1(t) = x_2(t) \\ \dot{x}_2(t) = -\frac{k}{M}x_1(t) - \frac{b}{M}x_2(t) + \frac{1}{M}F(t) \end{cases} \tag{6.5}$$

This is the state-space representation for the spring-mass model. Notice that the differential equation is second-order (split into two first-order equations), linear, and time-invariant (constant-coefficient).

It is important to keep in mind some of the assumptions and approximations used in this model before applying it. For example, the spring and dampening elements are assumed to be linear. The mass is approximated as a "point mass", regardless of the actual object's size and shape. Generally, this is fine for regular objects, but irregularly shaped objects where mass is not evenly distributed may cause significant errors, requiring a new or modified model.

6.11   Example 2: A Predator-Prey System

For a more intuitive mathematical modeling example, we consider the case of a predator-prey model. Here, a hypothetical population of rabbits and foxes are examined. The rabbit population grows based on a certain birth rate and (generally) decreases when eaten by predators (foxes). The fox population requires rabbits to prey upon as food in order to grow and a lack of food or a general death rate to shrink. The following elements of the system can be defined:

- x is the number of prey

- y is the number of predators

- $\frac{dx}{dt}$ is the growth of the prey population

- $\frac{dy}{dt}$ is the growth of the predator population

- α is the birthrate of the prey

- β is the death rate of the prey

- γ is the death rate of the predators

- δ is the birthrate of the predators

Additionally, several assumptions are made about the system for ease of generalization and conceptualization of the two groups:

- The prey always has enough food

- Predator food is entirely dependent upon their prey

- The rate of change of the population is proportional to its size

- Environmental factors do not significantly influence one group over the other

- Predators will always eat when possible

The prey population changes based on its birth rate (increasing the population) and its interaction with predators (decreasing the population when rabbits are eaten). It can be expressed mathematically by

$$\frac{dx}{dt} = \alpha x - \beta xy \qquad (6.6)$$

Similarly, the predator population increases when they are well-fed via the rabbit prey and decreases based on a natural death rate. As such, the predator group can be mathematically expressed as

$$\frac{dy}{dt} = \delta xy - \gamma y \qquad (6.7)$$

These two expressions can be combined to model the "dynamics" of the population pairs.

$$\frac{dx}{dt} = \alpha x - \beta xy$$
$$\frac{dy}{dt} = \delta xy - \gamma y \qquad (6.8)$$

Further, these equations can be represented in state-space form as follows:

$$\begin{cases} \dot{x} = \alpha x - \beta xy \\ \dot{y} = \delta xy - \gamma y \end{cases} \qquad (6.9)$$

The described predator–prey model is well-known and was developed by renowned researchers Lotka and Volterra. The numerical simulation of the system is shown in Figure 6.4. Notice that the rabbit and fox populations wax and wane cyclically and are slightly phase shifted. In other words, there are times when foxes eat too many rabbits, and there is subsequently not enough food, and foxes see a population decline. With fewer foxes, the rabbit population is able to increase, and the cycle begins anew.

The model has been used extensively to capture real-world phenomenon in various fields. For instance, predator–prey equations were used to build dynamic models to capture the interdependent behavior of transportation, economic, and environmental systems [51].

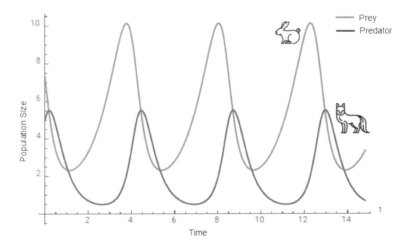

FIGURE 6.4: A predator-prey system.

6.12 Example 3: An RLC Circuit

Consider an RLC circuit as shown in Figure 6.5, where a current source is connected to a resistor, an inductor, and a capacitor in parallel. Given the input current, we are interested in finding the voltage output across the capacitor. Again, after drawing the basic diagram of the system we wish to model, we define our elements:

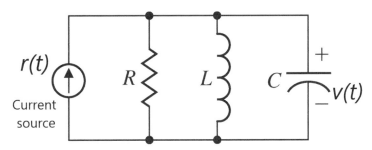

FIGURE 6.5: An RLC circuit.

- $r(t)$ is the current source
- R is the value of the resistor
- L is the value of the inductor

- C is the value of the capacitor

- $v(t)$ is the output voltage across the capacitor

Each element interacts with the current being applied and ultimately influences the final output voltage reading. Using Kirchoff's current law (an existing model of a physical law), we can express the model mathematically, as follows:

$$r(t) = \frac{v(t)}{R} + C\frac{dv}{dt} + \frac{1}{L}\int_0^t v(t)dt \qquad (6.10)$$

We could proceed as before to create a state-space version of the model. But, suppose we are not interested in a circuit model, and instead wish to model a mechanical system. In that case, we shall use the RLC circuit as an analog mathematical model to represent a mechanical system. In this model, the inductance L translates to mechanical system mass, the resistance R translates to a frictional coefficient, the reciprocal of capacitance $\frac{1}{C}$ translates to the spring constant, the voltage $v(t)$ translates to the velocity, and the current $r(t)$ translates to the external force. The analogous quantities of electrical and translational mechanical systems are summarized in Table 6.2.

TABLE 6.2: Electrical and transitional mechanical systems

Electrical System	Mechanical System
Voltage (V)	Force (F)
Resistance (R)	Frictional Coefficient (b)
Inductance (L)	Mass (M)
Reciprocal of Capacitance ($\frac{1}{C}$)	Spring Constant (k)
Current (i)	Velocity (v)
Charge (q)	Displacement (x)

Rewriting the electrical model in terms of velocity ($v(t)$), the differential equation of an RLC circuit (as an analog mathematical model for a mechanical system) becomes

$$M\dot{v}(t) + bv(t) + k\int_0^t v(t) = r(t) \qquad (6.11)$$

If we compare Equation 6.10 and Equation 6.11 and note the similarities, we can see how these are analogous mathematical models.

6.13   Example 4: An Epidemic Model

Now, consider a disease epidemic. Here, we define three category groups in a given population: susceptible, infected, and recovered individuals. Susceptible people are those that can be infected by the epidemic. Initially, nearly everyone is susceptible. Infected individuals have contracted the disease and are actively spreading it to susceptible individuals. Those who are recovered have been exposed to the disease already. But they may have built up an immunity to any further disease infection due to development of antibodies. The speed and strength with which the epidemic spreads, as well as how quickly people recover from the disease must also be considered. Based on these observations, the elements of the model are defined as follows:

- $s(t)$ is the group susceptible to infection

- $i(t)$ is the group currently infected

- $r(t)$ is the group who has recovered from the epidemic and is now immune

- β is the infection constant

- γ is the recovery constant

Let's examine the interactions of these three groups in order to build a mathematical model of the system. A change in the susceptible population will happen when that group interacts with a member of the infected group, resulting in a reduction in the susceptible population. This reduction is influenced by how quickly and easily the particular disease strain spreads. Likewise, the same interaction results in an equivalent increase in the population of infected individuals. Simultaneously, the population of infected individuals is decreasing due to members recovering from the disease and gaining immunity from further infection. Again, the recovery will happen at a certain strength and rate based on the particular disease strand. Finally, the recovered group will only increase as more and more individuals are infected by the epidemic and gain immunity. Given these relationships, the following set of state-space equations are constructed, representing each population group's change as the system evolves:

$$\begin{cases} \dot{s}(t) = -\beta s(t)i(t) \\ \dot{i}(t) = \beta s(t)i(t) - \gamma i(t) \\ \dot{r}(t) = \gamma i(t) \end{cases} \qquad (6.12)$$

A sample evolution of the epidemic model is shown in Figure 6.6. The system begins with nearly every individual as susceptible to the epidemic. As the system evolves, more and more become infected (hence fewer susceptible) and also some recover. Not all of the susceptible population will become infected unless there is a particularly aggressive and contagious disease. Eventually, all infected individuals will gain immunity from the disease and the becoming recovered (effectively ending the epidemic) as the system reaches an equilibrium point.

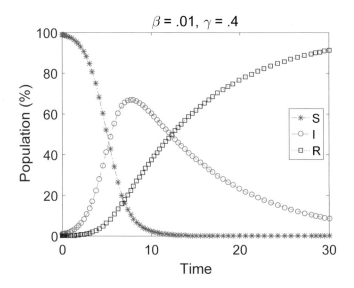

FIGURE 6.6: Sample evolution of an epidemic model.

It is important to note that this is an example of a "balanced" system. That is, all three groups are part of the same overall population that cannot grow or shrink. For example, when one individual becomes infected, they are added to the infected group and removed from the susceptible group. Mathematically, the relationship between the groups can be expressed as:

$$\dot{s}(t) + \dot{i}(t) + \dot{r}(t) = 0 \qquad (6.13)$$

Regardless of the distribution of the entire population among these three groups, its total does not change. Balanced systems may allow for some simplification to help in their mathematical solution.

6.14   Example 5: Vehicular Traffic Modeling

For a final example, we will try to model how traffic flows along a unidirectional road. Consider Figure 6.7, which illustrates a freeway system that can be controlled by ramp metering to affect traffic conditions on the freeway.

Traffic on the road can either be modeled microscopically by analyzing the behavior of individual cars or macroscopically by understanding the complex interaction of many vehicles on the road [53] [54]. We will keep the focus on macroscopic behavior and modeling of vehicles on the road. The three fundamental traffic variables are velocity, density, and flow. These variables are closely related, and their relationship is explained in the next section.

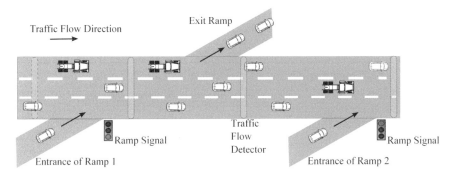

FIGURE 6.7: Freeway system and ramp metering. Courtesy of [52].

6.14.1 LWR and Greenshields' Models for Traffic

The macroscopic traffic flow model formulates the relationship among the key traffic flow parameters. The famous LWR (Lighthill-Whitham-Richards) model was proposed in 1956. It is a one-dimensional macroscopic traffic model named after the authors [55] [56]. The following equation gives the dynamics of traffic flow using the LWR model:

$$\frac{\partial}{\partial t}\rho(t,x) + \frac{\partial}{\partial x}f(t,x) = 0 \tag{6.14}$$

where ρ is the traffic density and f is the traffic flux or flow. Traffic flux is defined as the product of traffic density and the traffic speed v , i.e. $f = \rho \times v$. It is a partial differential equation formulating a relationship between traffic density and traffic flow using the conservation law — i.e., cars are neither created nor destroyed. This relationship is valid at every x and every t.

Many models link traffic density to traffic speed. One of them is Greenshields' model which proposes a linear relationship between traffic density and traffic speed [57]. This model is given as:

$$v(\rho) = v_f \left(1 - \frac{\rho}{\rho_m}\right) \tag{6.15}$$

where v_f is the free-flow speed and ρ_m is the maximum attainable density or jam density. Free-flow speed is the traffic speed when there is no traffic, i.e., when the traffic density is zero. Traffic jam density is the density at which there is a traffic jam, i.e., when the traffic speed is zero.

Traffic flow using Greenshields' model is given by:

$$f(t) = v_f\rho(t) \left(1 - \frac{\rho(t)}{\rho_m}\right) \tag{6.16}$$

To illustrate further, consider a case study using data from the Las Vegas freeway network. The objective of the case study was to gain deeper insights into the relationship among various traffic variables and also among the macroscopically aggregated variables. Data was collected by loop detectors placed every one-third of a mile on the I-15 northbound stretch. Data was aggregated at every 15-minute interval. Data used in our analysis was for one full year, starting from November 1st, 2014 to October 31st, 2015. Figure 6.8 shows the relationship between traffic variables flow and occupancy (also a measure of density) on all lanes at detector ID 112-2 when measured on Tuesdays. Figure 6.9 shows the relationship between traffic variables speed and occupancy on all lanes at detector ID 112-2, also measured on Tuesdays.

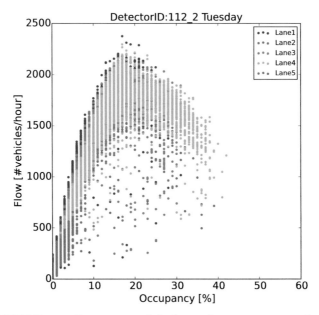

FIGURE 6.8: Experimental findings: flow-occupancy relationship.

6.14.2   ODE Approximation of LWR Model

Figure 6.11 illustrates space discretization of Equation (6.14) for the traffic flow on a road link. The ODE model for the traffic flow on the link is

$$\frac{d\rho(t)}{dt} = \frac{f_{in}(t) - f_{out}(t)}{\ell}, \tag{6.17}$$

where ℓ is the length of the network link, and f_{in} is the traffic inflow at the section. The outflow traffic using Greenshields' model is given by

$$f_{out}(t) = v_f \rho(t)\left(1 - \frac{\rho(t)}{\rho_m}\right). \tag{6.18}$$

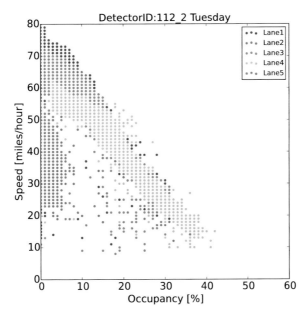

FIGURE 6.9: Experimental findings: speed-occupancy relationship.

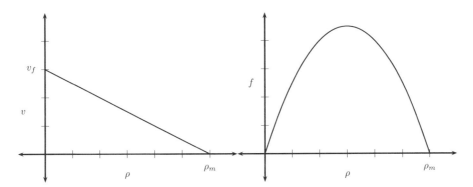

FIGURE 6.10: Fundamental diagram using Greenshields' model.

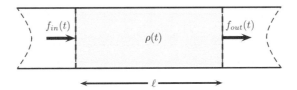

FIGURE 6.11: ODE model for traffic flow on a network arc. Courtesy of [58].

Diffusion is a handy concept mentioned in the literature of traffic flow models by many researchers. The introduction of the diffusion concept in traffic flow models makes them more realistic. The "diffusion effect" accounts for the fact that each driver is observing the road ahead of him and constantly adjusting his speed according to road density and flow condition. This adjustment makes the flow dependent on the gradient of density, leading to an effective diffusion. The diffusive term helps model the speed reductions due to shock waves as gradual ones rather than the sudden ones. Incorporating the diffusion term in Greenshields' model Equation (6.15) gives

$$v(\rho) = v_f \left(1 - \frac{\rho}{\rho_m}\right) - D \left(\frac{\partial \rho}{\partial x}\right) \Big/ \rho \qquad (6.19)$$

and we can rewrite the traffic flow using Equation (6.15) and (6.19) as

$$f(t) = v_f \rho(t) \left(1 - \frac{\rho(t)}{\rho_m}\right) - D \left(\frac{\partial \rho(t,x)}{\partial x}\right) \qquad (6.20)$$

where D is the diffusion coefficient.

Combining Equation (6.15) and (6.20) we get the distributed-diffusive model for traffic flow as follows

$$\frac{\partial}{\partial t}\rho(t,x) + v_f \frac{\partial}{\partial x}\rho(t,x) - 2\frac{\rho}{\rho_m}v_f \frac{\partial}{\partial x}\rho(t,x) - D\frac{\partial^2}{\partial x^2}\rho(t,x) = 0. \qquad (6.21)$$

6.15   Exercises

1. Discuss the significance of a mathematical model in science and engineering.

2. Discuss the main steps and challenges involved in developing a working model of a physical system.

3. You are tasked to design an adaptive traffic signal at a road intersection. Which modeling technique - micro or macro - would you prefer to use to obtain the solution. Discuss. Assume that you have the necessary sensors and historical data.

4. Why is model validation necessary?

5. Choose a current issue or discussion happening on social media. What kind of basic framework would you use to model it? Why? Identify the variables and parameters and explain what each represents in the case you chose.

6. A second-order ODE model for a spring-mass system can be obtained as shown in Equation 6.4. Express this equation in state-variable form.

7. For a simple pendulum shown in the figure below, the nonlinear equations of motion are given by

$$\ddot{\theta} + \frac{g}{L}sin\theta + \frac{k}{m}\dot{\theta} = 0$$

where g is gravity, L is the length of the pendulum, m is the mass attached at the end of the pendulum (assume a massless rod), and k is the coefficient of friction at the pivot point.

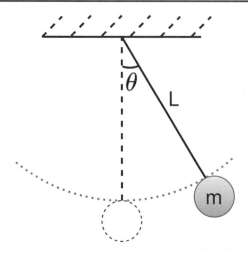

(a) Obtain a state variable (state space) representation of the system. The system output is the angle θ.

(b) Now, if the operation of the pendulum is about the equilibrium point, which is $\theta = 0$ in this case, you can use the approximation $sin\theta \approx \theta$. This will linearize the equations of motion about the equilibrium point. Now present this linear version of dynamics in the matrix form.

8. Consider the following mechanical system

$$\dot{X}(t) = \begin{bmatrix} 0 & 1 \\ -2 & -3 \end{bmatrix} X(t) + \begin{bmatrix} 0 \\ 1 \end{bmatrix} u(t)$$

$$y(t) = [1 \quad 0]X(t)$$

Using the lsim and ss commands in MATLAB simulate the system and obtain the (time) response of linear system. I.e. Obtain and plot the system responses (x_1 and x_2) and output response $y(t)$ when $u(t) = 0$.

9. The Van Der Pol oscillator is a non-conservative oscillator with non-linear damping. It evolves in time according to the second-order differential equation:

$$\frac{d^2x(t)}{dt^2} - \mu(1 - x^2(t))\frac{dx(t)}{dt} + x(t) = 0$$

where $x(t)$ is the position coordinate which is a function of the time

t, and μ is a scalar parameter indicating the nonlinearity and the strength of the damping.

(a) Find the state space representation of the system

(b) Is the system linear? If yes, present the system in a matrix form.

10. Use ODE45 to numerically solve the nonlinear ODE (Van Der Pol) as described in previous question. [t,y] = ode45(odefun,tspan,y0), where tspan = $[t_0 \ t_f]$, integrates the system of differential equations $y\prime = f(t, y)$ from t_0 to t_f with initial conditions y_0. You may use $\mu = 1$, initial conditions $x_1(0) = 1$ and $x_2(0) = 1$, and tspan = [0 20]. Your solution should contain three plots — system responses x_1 and x_2 with respect to time and phase portrait, i.e. x_1 with respect to x_2.

7

Epidemiology-Based Models for Information Spread

> To understand the Universe, you must understand the language in which it's written, the language of Mathematics.

Galileo Galilei

With a basic understanding of models and how models are developed, this chapter examines epidemiology models used to describe how disease is spread throughout a population, including the susceptible-infectious-recovered (SIR) and susceptible-exposed-infectious-recovered (SEIR) disease spread models. The concepts of herd immunity and curve flattening are explained in the context of these epidemic models. These models are then framed as analog models that lead to several well-established information spread models based on epidemiological principles. After highlighting some fundamental terminology, the Ignorant-Spreader (IS), Ignorant-Spreader-Ignorant (ISI), and Ignorant-Spreader-Recovered (ISR) models for information spread are examined, as well as the herd immunity concept translated to immunity from information epidemics. Using these models and social network theory, several modified epidemiology-based models are presented to address the unique assumptions and considerations present in online social media information spread that are not accounted for in classic ISR-type person-to-person information spread models.

7.1 Epidemiology Models

This section will expand upon the epidemic model example from the previous chapter and offer new insight into modeling epidemiology-based models. Additionally, we will draw a connection between classic epidemiology models and information spread models. From the previous chapter, it is reasonably intuitive to see state changes (or transitions) as a set of directional inputs and outputs from that state. Consider a basic state transition from one state to another, as shown in Figure 7.1. Everything entering a state "box" is added to that state, and terms leaving a state box are subtracted from that state. These state boxes can be used intuitively to build dynamic models in state-space representation.

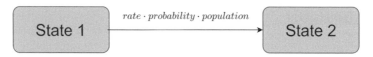

FIGURE 7.1: A state transition for an epidemic model.

Generally speaking, for epidemic models, these states represent the number of people in each group or class. In the case of epidemic models, the number of people transitioning to another group is determined by the rate at which states change, the probability it will change for any given interaction, and the population of people being considered. As one state gains a portion of the population in a given time, the other will lose it.

7.1.1   The SIR Disease Spread Model

Recall that for an SIR-based epidemiology model, there are three main states: susceptible, infectious, and recovered. People who are susceptible can potentially contract a disease and become infectious. These infectious people will spread the disease throughout the remaining susceptible people. As time passes and people begin to recover from the disease, they gain immunity to further infection. As a result, infectious people are coming into contact with both the remaining susceptible people and those who have recovered. Since these interactions are essentially random, an infectious contact with someone who has recovered is a contact that will not lead to a new infection. It effectively slows down the spread of the disease until enough growth in the number of recovered people halts most of the disease's progression and makes it unlikely that it will live long enough to spread to the remaining susceptible people.

In addition to the model variables, there are several parameters of concern: the spreading rate, recovery rate, connectedness, and population size. The spreading rate β is the rate at which infectious people spread the disease to susceptible people. Likewise, the recovery rate γ represents how quickly infectious people recover from the disease and gain immunity. The average connectedness of people within the population is denoted by k. A high connectedness value means that the population is highly interconnected, and any given person will be able to make direct contact with most of the group. In contrast, low connectivity values represent reasonably sparse networks with fewer direct connections to others in the population. Refer to Chapter 4 and Chapter 5 for details on networks and connectedness. Obviously, higher values of k will lead to a much more rapidly spreading (but also rapidly recoverable) disease. The population size of N is simply the number of people in the population being examined. It is common to let $N = 1$ or $N = 100$ when one either doesn't care about the population size or is only interested in the percentage of the population represented in each state class. The SIR model variables and parameters are summarized in Table 7.1.

TABLE 7.1: SIR model variables and parameters.

Term	Meaning
$S(t)$	Susceptible: people potentially able to be infected at time t
$I(t)$	Infectious: people spreading the infection at time t
$R(t)$	Recovered: people no longer infected and immune at time t
β	Spreading rate
γ	Recovery rate
k	Average connectedness of individuals
N	Population size
R_0	Number of people an infectious individual infects ($\frac{\beta}{\gamma}$)

It should be noted that several assumptions are used in SIR-based epidemic models. These assumptions are often used to simplify the models enough to keep them generalized, and to address factors that may not be observable or are burdensome to determine with a good degree of certainty. The following assumptions are made [59]:

1. The population has a constant size of N. Because it is generally a large number, classes can be considered as continuous variables. In addition, birth and natural death rates are equal.

2. The population is mixed homogeneously. Specifically, equal expected value and variance is assumed for each variable within the population. One sampling can represent the population on average, even if small localized differences exist in reality.

3. People recover and are removed from the infectious class at a rate proportional to the number of those infected. It is known as the daily recovery removal rate.

For SIR-based models, recovery grants immunity from further infection. It does not hold for all epidemic models (such as SIS models). SIR-based models are appropriate for many viral afflictions such as measles, mumps, and smallpox.

With variables, parameters, and assumptions addressed, a block diagram for the SIR model's state transitions can be created, as shown in Figure 7.2.

The transition from the susceptible to the infectious state involves the interaction of susceptible people with infectious people at an infection rate of β with population N and connectedness of k. Likewise, the transition from

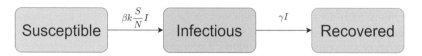

FIGURE 7.2: State transitions for an SIR epidemic model.

the infectious to the recovered state occurs when infectious people recover at a rate of γ.

With a directional block diagram model of the SIR system and knowledge of each state transition properties, the system dynamics mathematical model can be developed. In state-space representation, the system dynamics are expressed as:

$$
\begin{cases}
\dot{s}(t) = -\frac{\beta k}{N} s(t) i(t) \\
\dot{i}(t) = \frac{\beta k}{N} s(t) i(t) - \gamma i(t) \ . \\
\dot{r}(t) = \gamma i(t)
\end{cases}
\tag{7.1}
$$

Note that the dynamics are reasonably intuitive when considering the input-output interactions and the associated parameters for each state box transition pair. A sample simulation plot of a SIR epidemic model is given in Figure 7.3. In this example, not all of the susceptible people are infected, and the infectious population is kept relatively low. After time progresses sufficiently, every infectious person eventually recovers, and the susceptible and recovered population curves are mirrored.

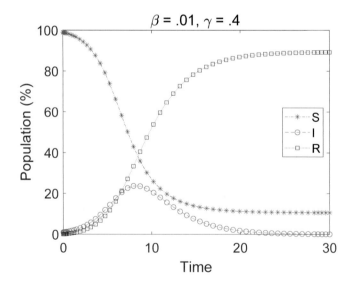

FIGURE 7.3: Sample system evolution of an SIR epidemic model.

Since we are assuming a population of a constant size N, and we know that the population consists of some distribution of susceptible, infectious, and recovered people at any given time, N can be expressed as a sum of the three states as shown in Figure 7.2. Because the total population is not changing, anyone removed from one state is moved to another, and nobody is entering or leaving the population in aggregate. It is known as a "balanced" system. Because of this balanced relationship and the existence of a fourth equation to relate the states to each other, only two of the three primary system dynamics equations are needed to examine the entire system.

$$\dot{s}(t) + \dot{i}(t) + \dot{r}(t) = 0 . \tag{7.2}$$

The basic reproduction number R_0 is a useful metric to determine the "infectiousness" of a disease. It is defined as the expected number of new infections that come from a single infectious person in a population of fully susceptible people. In the case of a basic SIR-type model, it is the ratio of the rate of infection β and the rate of recovery γ, as shown in Equation 7.3. Because R_0 is a ratio of rates, it is dimensionless. High values of R_0 signify an exceptionally infectious disease, while low R_0 values are reasonably mild.

$$R_0 = \frac{\beta}{\gamma} . \tag{7.3}$$

Next, we will examine a slightly more complicated SIR-based epidemiology model. Despite the added complexity, it still follows the same fundamental framework as the basic SIR model.

7.1.2 The SEIR Disease Spread Model

If the incubation period for a disease is especially long, the basic SIR model will become increasingly less effective at modeling the actual behavior of a disease progression as it spreads through a population. The SEIR model modifies the basic SIR model to include exposed people as a new state class. People in the exposed state have contracted the disease, but are not yet infectious. Assumptions for SEIR models are similar to those of basic SIR models. Figure 7.4 shows a block diagram for SEIR model state transitions.

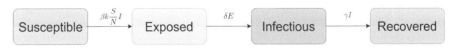

FIGURE 7.4: State transitions for an SEIR epidemic model.

Instead of going from a susceptible to an infectious state, people first transition to the exposed state using the same mathematical transition properties as the SIR model's susceptible to infectious transition. The probability of infection, rate of infection, and the population remain identical. Similarly, the infectious to recovered state transition is similar to the previous model, leaving exposed to infectious transition to be determined. Because everyone who is exposed eventually becomes infected, the probability of infection is 1. The population of those exposed is defined as E, and the rate of exposed people becoming infectious is set as δ. The system dynamics shown in Equation 7.4 can be obtained using the same methods as described using the basic SIR model.

$$\begin{cases} \dot{s}(t) = -\frac{\beta k}{N} s(t) i(t) \\ \dot{e}(t) = \frac{\beta k}{N} s(t) i(t) \\ \dot{i}(t) = \delta e(t) - \gamma i(t) \\ \dot{r}(t) = \gamma i(t) \end{cases} . \tag{7.4}$$

Like the basic SIR model and its underlying assumptions, the SEIR epidemic model is a balanced system, shown in Equation 7.5. The exposed group of people is simply another subgroup of the population N. In this case, three equations are required to describe the system entirely, as

$$\dot{s}(t) + \dot{e}(t) + \dot{i}(t) + \dot{r}(t) = 0 . \tag{7.5}$$

Equation 7.6 gives the basic reproduction number R_0 for the SEIR epidemic model, where δ is the incubation period, and μ is the death rate. The proof for finding R_0, in this case, is beyond this text's scope, but the concept of the anticipated number of infections that come from one infectious person remains the same.

$$R_0 = \frac{\delta}{\mu + \delta} \frac{\beta}{\mu + \gamma} . \tag{7.6}$$

7.1.3   Herd Immunity in Epidemiology

Can a population eventually gain immunity to an infectious disease? Perhaps. The answer largely depends on the type of infectious disease and if a vaccine can be developed. In SIR-type illnesses, people can recover and gain immunity via the development of antibodies, which in turn can be used to develop vaccines. Other infectious diseases such as SIS-based diseases allow "recovered" individuals to become re-infected after a time, and vaccines cannot be effectively designed. For now, however, we will focus on SIR-based infectious diseases.

If a population manages to largely recover from a disease and build antibodies (or get vaccinated), the recovered segment of the population can provide indirect protection to the population's remaining susceptible portion. This concept is known as "herd immunity" or herd protection. Consider Figure 7.5 and note the difference in infectious and susceptible individuals as an increasing amount of recovered or vaccinated people are present. Assuming each contagious person can, on average, infect up to two other people, the number of infectious people will increase unless there are recovered or vaccinated people to "block" further spread to susceptible people.

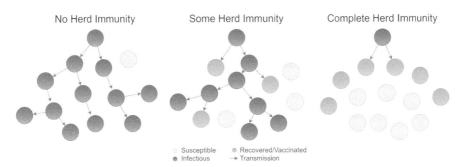

FIGURE 7.5: Herd immunity in SIR-based epidemic models.

Infectious diseases, such as measles, mumps, polio, chickenpox, and more were common diseases in the United States, but are now relatively rare. It is mainly due to the effects of herd immunity [60]. Since these diseases are no longer actively spreading, herd immunity is obtained through vaccinations (typically administered at a young age before beginning school). Sometimes, outbreaks of herd immunity mitigated infectious diseases can still be felt in communities with lower vaccine coverage (reducing the benefit of herd immunity for the entire susceptible population segment). One example of such an outbreak was in 2019 at Disneyland and Universal Studios in Southern California. An outbreak of measles (one of the most contagious viruses in the world) was spread by an infectious unvaccinated teen who visited the parks [61].

7.1.4 "Flattening the Curve"

Flattening the curve refers to community isolation measures or stay at home orders intended to mitigate the number of disease cases per day to a manageable level so that medical practitioners and medical resources (hospital beds, limited specialized equipment, etc.) do not become overwhelmed. Figure 7.6 shows a SIR curve with both no intervention and the proactive application of mitigation efforts.

There are two main ways in which an epidemic curve is flattened. The first method is by lowering the connectedness factor, k. In other words, lessen the strength of direct contact between individuals within a population. It is usually achieved through "social distancing" measures meant to keep people from close physical proximity to each other or reduce the number of people allowed to patronize a business simultaneously. The second method is to reduce the number of susceptible people within a group. This can be accomplished through extreme lock-down orders or vaccinations (if available). By reducing the number of susceptible people, fewer can become infectious at once, allowing time for recovery and a gradual move toward herd immunity protection. It was observed worldwide during the COVID-19 outbreak in 2020.

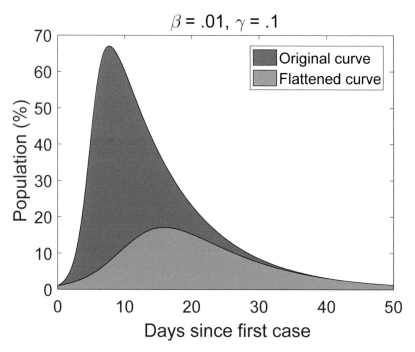

FIGURE 7.6: Comparing no intervention and intervention strategies.

```
Code 7.1: Comparing Original and Flattened Curves
1  clear;
2  orange = [0.9290, 0.6940, 0.1250]; %customized
       color
3  to = 0; %start time of original curve
4  tf =50; %final time of original curve
5  yo = [99 1 0]; %original susceptible population
6  to2 = 0; %start time of original curve
7  tf2 = 50; %start time of original curve
8  yo2 = [40 1 0]; %adjusted susceptible population
9  [t, y] = ode45('ypSIR_CurveFlattening',[to tf],yo)
       ; %calls ode45
10 [t2, y2] = ode45('ypSIR_CurveFlattening',[to2 tf2
       ],yo2);
11 cla
```

```
12  area(t, y(:,2), 'FaceColor','m') %plots original
       curve
13  hold on
14  area(t2, y2(:,2), 'FaceColor',orange) %plots
       flattened curve
15  hold on
16  title('\beta = .01, \gamma = .1', 'FontWeight','
       Normal')
17  xlabel('Days since first case')
18  ylabel('Population (%)')
19  legend('Original curve','Flattened curve')
20  ax = gca;
21  ax.FontSize = 16;
```

```
1   function ypsir_basic = ypSIR_CurveFlattening(t,y)
2   %sent to ode45 to calculate SIR dynamics (must
       supply infection and
3   %recovery rates)
4   a = .01;
5   b = .1;
6   ypsir_basic(1) =-a*y(1)*y(2);
7   ypsir_basic(2) = a*y(1)*y(2)-b*y(2);
8   ypsir_basic(3) = b*y(2);
9   ypsir_basic = [ypsir_basic(1) ypsir_basic(2)
       ypsir_basic(3)]';
10
11  end
```

7.1.5 Epidemiology Models as Analog Models

So why present epidemiology models when our primary interest is information spread? Recall that in the previous chapter, we discussed some popular types of models. In a large portion of this chapter, we will be using epidemiology models as an analogous model for information spread modeling. Indeed, several model modifications or adjustments must be made to these models in a context-sensitive way to apply them effectively to information spread. However, the underlying process of people being susceptible to new information, spreading it to others like an infection, and recovering from it by losing interest mirrors several well-established existing epidemiology models closely.

Just like with epidemiology models, assumptions and simplifications can help make information spread models more manageable and practical. Although N is, in reality, a variable, it is assumed to be constant for simplicity. For mathematical simplification, the population can be viewed as a mixed group of states, each as a percentage of the entire population when $N = 1$. The connectedness factor k is often omitted when average connectedness is either assumed to be very high or not relevant to the system being described. Online connectedness, for example, will be very high compared to traditional SIR-based model assumptions. Each of us is likely connected to far more people online while browsing Twitter, Facebook, and Instagram than in person-to-person encounters with other individuals.

7.2 Information Spread Models: Overview and Conventions

Many deterministic models focus on a system's mathematical dynamics. That is, it describes the time dependence of a point in relation to its position (though it need not be a physical position). Examples of this can include a simple system like a swinging pendulum or complex systems such as the traffic flow on a highway [62]. In deterministic dynamical models, only one future state can follow from the current state. Many basic deterministic information spread models can trace their dynamical origins to special applications of epidemiology-based disease spread models. As such, much of the epidemiology terminology requires modification for ease of use and understanding in the context of information spread. The fundamental variables and parameters used here are summarized in Table 7.2. Additionally, since information spread applications are the focus of this book, some common nomenclature used to reference traditional epidemic model elements are altered to properly reflect their context-specific usage. Specifically, epidemic spreaders and infected classes have been renamed as ignorants and spreaders, respectively.

TABLE 7.2: Essential variables and parameters.

Term	Meaning
I	Ignorants: class which does not know the information
S	Spreaders: class which is spreading the information
R	Recovered: class which no longer spreads the information
β	Information spreading rate
γ	Information recovery rate
k	Average connectedness of individuals

In the simple deterministic models presented, several assumptions are made as follows:

1. Information is transmitted via spreader contact with an ignorant individual.

2. Upon contact, information transmission is instantaneous.

3. All ignorant and spreading individuals are equally susceptible and trans-mittable, respectively.

4. The population size is fixed.

The rest of this chapter discusses several established models and their modifications. The models, their applications, and a few examples are presented.

7.3 The Ignorant-Spreader Model

Consider the most basic information spreading scenario: everyone in the population is either unaware (ignorant) of the message or knows it and spreads it around to other community members. The flow diagram for the IS model is shown in Figure 7.7. Intuitively, this model's applications may not be immediately evident when first examined, as one would envision individuals lose enthusiasm for spreading a specific snippet of data or news over time. While this is ultimately typically true, over a specified span of time, information can spread without stopping. For example, long-term militant social or cultural movements or important news that everyone would care about during a brief time-span, like a local natural disaster alert, exhibit this kind of information spreading behavior. For the purposes here, one can think of the IS model as a subset of the ISR model discussed in later sections, but with limitations. In the IS model, ignorants are always decreasing while spreaders are increasing as the information spreads throughout the population.

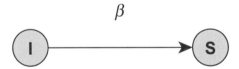

FIGURE 7.7: Flow diagram for the Ignorant-Spreader model.

The system dynamics of the IS model are expressed mathematically as follows:

$$\begin{cases} \dot{i}(t) = -\beta k i(t) s(t) \\ \dot{s}(t) = \beta k i(t) s(t). \end{cases} \tag{7.7}$$

Simulating the system dynamics under typical parameters results in Figure 7.8. It is obvious that the spreaders eventually overtake the entire population, at which point everyone is aware of the information and actively spreading it.

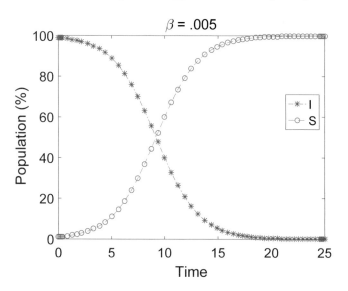

FIGURE 7.8: Time Evolution of an Ignorant-Spreader model

Code 7.2: Ignorant-Spreader Model

```
clear;
to = 0; % initial time
tf =25; % final time
yo = [99 1]; % initial ignorants and spreaders
[t y] = ode45('ypsi',[to tf],yo); % call to ode45
    and plotting the results
plot(t,y(:,1),'-.r*',t,y(:,2),'--mo')
title('\beta = .005', 'FontWeight','Normal')
xlabel('Time')
ylabel('Population (%)')
legend('I','S','Location','Best')
ax = gca;
ax.FontSize = 16;
```

```
function ypsi = ypsi(t,y)
% Sets up the dynamics to be called to ode45
b = .005;   %spreading rate, beta
ypsi(1) =-b*y(1)*y(2);
ypsi(2) = b*y(1)*y(2);
ypsi = [ypsi(1) ypsi(2)]';

end
```

7.4   The Ignorant-Spreader-Ignorant (ISI) Model

Using the Ignorant-Spreader model as a foundation, consider the case in which ignorants become spreaders as they learn information, just as before. However, these spreaders then recover from the information spreading state (due to disinterest over "old news," for example). Each recovered individual then immediately returns to a state where they can become re-exposed to the information, regain interest, and spread it again. Typical examples of this include new variations on the latest celebrity gossip or renewed interest in a popular technology product due to a new and exciting design. The ISI model flow diagram is outlined in Figure 7.9.

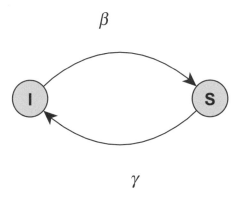

FIGURE 7.9: Flow diagram for the Ignorant-Spreader-Ignorant model.

The system dynamics of the ISI model are taken to be:

$$\begin{cases} \dot{i}(t) = -\beta k i(t) s(t) + \gamma s(t) \\ \dot{s}(t) = \beta k i(t) s(t) - \gamma s(t). \end{cases} \tag{7.8}$$

Notice that the main difference between the ISI model and the previously described IS model dynamics is the addition of a stifling parameter γ, representing the stagnating of spread information, reverting the spreaders to ignorant class for a specific topic or topic set. A simulation of an ISI model is shown in Figure 7.10.

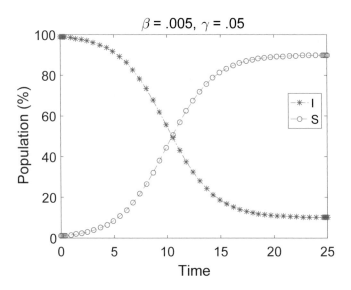

FIGURE 7.10: Sample evolution of the Ignorant-Spreader-Ignorant model.

Code 7.3: Ignorant-Spreader-Ignorant Model

```
1  %Graphs the time evolution of the dynamics
2  clear;
3  to = 0;
4  tf =25;
5  yo = [99 1];
6  [t y] = ode45('ypISS',[to tf],yo);
7  plot(t,y(:,1),'-.r*',t,y(:,2),'--mo')
8  title('\beta = .005, \gamma = .05', 'FontWeight','
       Normal')
9  xlabel('Time')
10 ylabel('Population (%)')
11 legend('I','S','Location','Best')
12 ax = gca;
13 ax.FontSize = 16;
```

```
1  function ypISS = ypISS(t,y)
2  %This function outputs the dynamics to be analyzed
3  a = .05;   %stifling constant
4  b = .005;    %spreading constant
5  k = 1;   %interaction constant
6  % Dynamics to be called to ode45
7  ypISS(1) =-b*k*y(1)*y(2)+a*k*y(2);
8  ypISS(2) = b*k*y(1)*y(2)-a*k*y(2);
9  ypISS = [ypISS(1) ypISS(2)]';
10 end
```

7.5 The Ignorant-Spreader-Recovered (ISR) Model

In many cases, a message that is learned and spread is eventually simply abandoned as enthusiasm for the subject degrades, and community members are not re-exposed to the same news, gossip, etc. The Ignorant-Spreader-Recovered model is often referred to as the Maki-Thomson model, and its flow diagram is given in Figure 7.11. Recovered individuals are former members of the spreader class that are disinterested in spreading the information, have tired of the topic, or are perhaps even aware of its true-false nature and see no point to spread it further. While in the recovered state, community members continue to communicate pairwise with other people in the network. Recovered individuals communicating with spreaders can sometimes signal to the spreader that the news being spread in that interaction is not worth spreading because it is "old news", hence changing the spreader to the recovered class. Likewise, two communicating spreaders can indicate that the information is already well-known and no longer requires spreading, transforming one of the pair into a recovered state. Naturally, the occurrence of class state changes following a pairwise interaction is probabilistic and dependent on the stifling rate of the news. These class interactions are summarized in Table 7.3. Examples of ISR types of information spread include fact-based news such as a natural disaster occurring or even broad public reactions to a highly anticipated television program episode. It should be noted that ISI types of information spread can be viewed in the ISR form in certain cases if one considers each new piece of information about a larger topic independently. For example, while the screen size of the latest cellphone product might be modeled as an ISR system with its information propagation, the product brand as a whole could follow the ISI model as interest is renewed with the latest brand iteration.

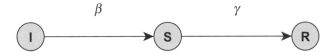

FIGURE 7.11: Flow diagram for the Ignorant-Spreader-Recovered model.

TABLE 7.3: ISR class interactions.

Interaction	Result
$I + S + R = 1$	Conservation of individuals in the population
$I + S \rightarrow 2S$	Spreader will infect an ignorant with the message
$S + S \rightarrow S + R$	One spreader will recover if two interact
$S + R \rightarrow 2R$	Spreader will recover if contacting a recovered

The system dynamics of the ISR model are as follows:

$$\begin{cases} \dot{i}(t) = -\beta k i(t) s(t) \\ \dot{s}(t) = \beta k i(t) s(t) - \gamma k s(t)[s(t) + r(t)] \\ \dot{r}(t) = \gamma k s(t)[s(t) + r(t)]. \end{cases} \tag{7.9}$$

Figure 7.12 shows simulation results for the ISR model and illustrates the evolution of the ignorant, spreader, and the recovered class.

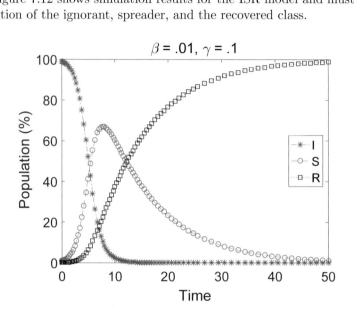

FIGURE 7.12: Sample time evolution of an ISR model.

```
              Code 7.4: Ignorant-Spreader-Recovered Model
 1  clear;
 2  to = 0;
 3  tf =50;
 4  yo = [99 1 0];
 5  [t y] = ode45('ypisr_basic',[to tf],yo);
 6  plot(t,y(:,1),'-.r*',t,y(:,2),'--mo',t,y(:,3),':bs
       ')
 7  title('\beta = .01, \gamma = .1', 'FontWeight','
       Normal')
 8  xlabel('Time')
 9  ylabel('Population (%)')
10  legend('I','S','R','Location','best')
11  ax = gca;
```

```
12  ax.FontSize = 16;
```

```
1   function ypsir_basic = ypisr_basic(t,y)
2   % Sets up the dynamics to be called to ode45
3   a = .01; % spreading rate
4   b = .1; % stifling rate
5   ypsir_basic(1) =-a*y(1)*y(2);
6   ypsir_basic(2) = a*y(1)*y(2)-b*y(2);
7   ypsir_basic(3) = b*y(2);
8   ypsir_basic = [ypsir_basic(1) ypsir_basic(2)
        ypsir_basic(3)]';
9
10  end
```

7.6 Reproductive Number and Herd Immunity

One helpful measurement for examining whether or not a message will spread through a populace or dissolve before any significant number of individuals become conscious of it is the basic reproductive number, denoted by R_0. It was initially utilized in its present day form by George MacDonald in 1952 to model the spread of malaria throughout a community. Epidemiology specific applications aside, it can be easily applied to an assortment of frameworks in which a growth and decay factor are in opposition with one another. Earlier, this concept was discussed in its original application to epidemiological systems. Here, we extend this concept to information spread interaction applications. Expressed simply, if the message is spreading with more "strength" than it is being silenced or abandoned, then the information takes on epidemic qualities and propagates throughout the community. Otherwise, the information will deteriorate over time. R_0 is a threshold condition. Formally, R_0 is still defined as the expected number of secondary cases that arise from a single spreader in a group that is otherwise ignorant of the information. R_0 is not a rate, but instead dimensionless. In general, the basic reproductive number in the context of information spread is defined as:

$$R_0 = \frac{k(\gamma + \beta)}{\gamma} > 1 \atop \frac{k\beta}{\gamma} > 0. \tag{7.10}$$

The numeric value of the basic reproduction number can substantially affect the degree to which information and news spreads, if it spreads at all. Compare the three sample ISR evolutions in Figures 7.13, 7.14, and 7.15. Each evolution has a different R_0 value. Note in Figure 7.14 the reproductive number of the information does not permit it to encounter the whole ignorant class, but in Figure 7.15 the information spreads to a high percentage of the community rapidly followed by a similarly quick recovery.

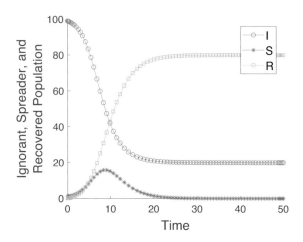

FIGURE 7.13: Sample time evolution of an ISR model.

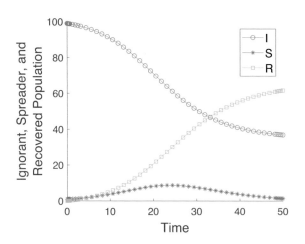

FIGURE 7.14: R_0 Comparison: low R_0 value.

Likewise, the epidemiological concept of herd immunity can viewed within the context of social media information spread. Conceptually, the process is similar. People who are spreading information or rumors will have less success and reach when those ignorant of it are recovered or immunized.

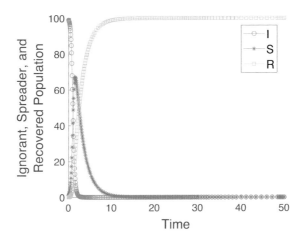

FIGURE 7.15: R_0 Comparison: high R_0 value.

For example, let's assume that a celebrity is rumored to have recently died in some type of accident. In reality, the celebrity is unharmed and, in fact, was never even involved in an accident. Regardless, news of the apparent death spreads throughout various tabloid magazines, internet blogs, and Twitter tweets. At first, most of those ignorant to the information have little reason to disbelieve the story. In time, however, it is discovered that the celebrity death declaration was untrue, but several sources of the initial rumor do not print retractions or fact updates. Luckily, tweets and more carefully fact-checking news sites report the correct information and educate both those who have heard the false rumor and also those who have not yet learned of the "death" in the first place. These two groups are effectively recovered and immunized from further information, respectively.

This conceptual method of translating epidemiological herd immunity concepts to information spread can be applied to many context-sensitive cases. That said, it is essential to remember that just like epidemiological SIR-based models, herd immunity can only occur if an immunity or vaccination analog is possible. Information that sees cyclical or renewed interest with new developments, for example, can never enter a real herd immunity state but may prove resistant to a specific iteration of information for a short time.

7.7 ISR Model for Social Media

Although the ISR information spread model functions well for "traditional" forms of pairwise interactions, context-aware assumptions and review must be made when it concerns current social media networks [63]. But why does digital social media require special consideration? Because person-to-person virtual interactions occur very rapidly and deliberately. When an individual is immediately notified of an inter-network social network post and decides to re-post or otherwise propagate the information, the spreading and stifling parameters can be very unpredictable and fluctuate wildly based on social impulses. Also, when a person in a social network has recovered from a particular news story or information item, that person is not positioning themselves in a position to knowingly and actively engage in pairwise contact with people from the spreader class (at least for that specific news). As a result, the spreader class is left unchecked from people in the recovered class and hence spreaders only realizing that the story they spread is broadly known upon contact with other spreaders. This modification to the standard ISR rumor model is new and distinct from the traditional model.

The resulting equations for an ISR system in social media are as follows [63]:

$$\left\{ \begin{array}{l} \dot{i}(t) = -\beta k i(t) s(t) \\ \dot{s}(t) = \beta k i(t) s(t) - \gamma k s(t)[s(t)] \\ \dot{r}(t) = \gamma k s(t)[s(t)]. \end{array} \right. \tag{7.11}$$

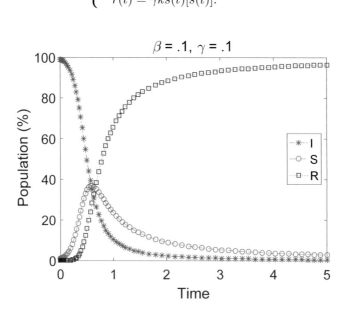

FIGURE 7.16: Sample time evolution of the ISR model for social media.

It is significant to note that in typical online social media communities, information spreads more rapidly (though not necessarily as widely) and decays at a different rate than traditional non-digital network rates, in part due to the loss of stifling factors from the recovered class. A sample simulation of an ISR model, modified for social media applications is shown in Figure 7.16.

Using the same parameters as in-person rumor spreading models, the information will not be as comparatively widespread. However, there will be far more interactions over the same amount of time which must then be measured in very different time scales. In a traditional rumor model, one might interact by chatting with a neighbor or reading a newspaper article over the course of a day. In an online social media environment, one could browse Twitter or Facebook for ten minutes and suddenly have hundreds of directional interactions simply by absorbing the posts of other users. Many of them will likely be ignored and not lead to widespread "sharing" (or spreading) of the information, but a select few of particular interest may be spread. Additionally, a few of the user's own thoughts, pictures, or memes may be shared on a social media site in those same ten minutes, adding to the sea of information waiting to be noticed and then ignored or spread.

Code 7.5: Ignorant-Spreader-Recovered Model for Social Media

```
%Graphs the time evolution of the dynamics
clear;
to = 0;
tf =5;
yo = [99 1 0];
[t y] = ode45('ypISR_sm',[to tf],yo);
plot(t,y(:,1),'-.r*',t,y(:,2),'--mo',t,y(:,3),':bs
   ')
title('\beta = .1, \gamma = .1', 'FontWeight','
   Normal')
xlabel('Time')
ylabel('Population (%)')
legend('I','S','R','Location','best')
ax = gca;
ax.FontSize = 16;
```

```
function ypISR_sm = ypISR_sm(t,y)
%This function outputs the dynamics to be analyzed
%   Detailed explanation goes here
a = .1; %.03;  %stifling constant
b = .1; %.6;   %spreading constant
k = 1; %.1;   %interaction constant
ypISR_sm(1) =-b*k*y(1)*y(2);
ypISR_sm(2) = b*k*y(1)*y(2)-a*k*y(2)*(y(2));
ypISR_sm(3) = a*k*y(2)*(y(2));
ypISR_sm = [ypISR_sm(1) ypISR_sm(2) ypISR_sm(3)]';

end
```

7.7.1   ISR Model for Social Media with Decay

Now we examine a scenario in which online social media information or news is dispersed over an extended time period. A message is learned, spread, and eventually recovered from, but people's natural boredom of an outdated story becomes a noteworthy factor. To account for this conduct, a basic exponential disinterest factor supplements the spreader dynamics. This method is much like a model element present in the Vidale-Wolfe advertising model [64], discussed in Chapter 9. The decay factor is offset by the dynamics of the recovered group, as those people who cease caring about a message over time essentially operate as recovered individuals.

The dynamics equations for a social media ISR system with decay are as follows:

$$\begin{cases} \dot{i}(t) = -\beta k i(t) s(t) \\ \dot{s}(t) = \beta k i(t) s(t) - \gamma k s(t)[s(t)] - \delta s(t) \\ \dot{r}(t) = \gamma k s(t)[s(t)] + \delta s(t). \end{cases} \tag{7.12}$$

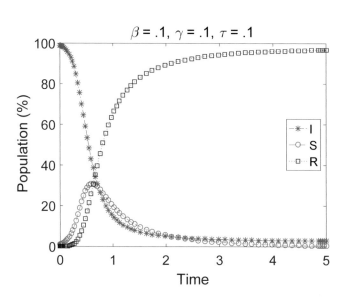

FIGURE 7.17: Sample time evolution of the ISR model for social media with decay.

Over long periods of digital media information spreading throughout a network, spreader decay reshapes spreader behavior to more closely resemble traditional ISR information spread curves. Figure 7.17 displays a sample evolution of a social media ISR model with an added decay factor. Obviously, the relevance and "juiciness" of the news or information will significantly influence the rate of natural spreading decay.

Code 7.6: Ignorant-Spreader-Recovered Model for Social Media with Decay

```
%Graphs the time evolution of the dynamics
clear;
to = 0;
tf =5;
yo = [99 1 0];
[t y] = ode45('ypISR_d',[to tf],yo);
plot(t,y(:,1),'-.r*',t,y(:,2),'--mo',t,y(:,3),':bs
    ')
title('\beta = .1, \gamma = .1, \tau = .1', '
    FontWeight','Normal')
xlabel('Time')
ylabel('Population (%)')
legend('I','S','R','Location','best')
ax = gca;
ax.FontSize = 16;
```

```
function ypISR_d = ypISR_d(t,y)
%This function outputs the dynamics to be analyzed
%    Detailed explanation goes here
a = .1;  %stifling constant
b = .1; %spreading constant
k = 1;   %interaction constant
tau = .1; %natural decay time constant
ypISR_d(1) =-b*k*y(1)*y(2);
ypISR_d(2) = b*k*y(1)*y(2)-a*k*y(2)*(y(2))-y(2)
    *(1-exp(-t/tau));
ypISR_d(3) = a*k*y(2)*(y(2))+y(2)*(1-exp(-t/tau));
ypISR_d = [ypISR_d(1) ypISR_d(2) ypISR_d(3)]';

end
```

7.8 ISCR Model for Contentious Information Spread

To examine a single community where "contentious information" is being dispersed, another novel model is introduced to represent spreaders and counter-spreaders of social media-based messages [63]. The basic ISR model performs well as a starting point for this proposed model but falls short when examining

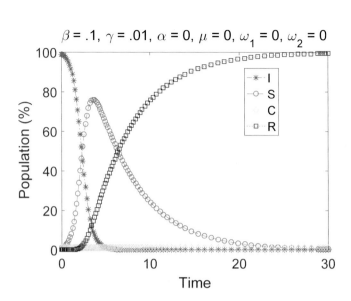

FIGURE 7.19: ISCR: No counter-spreaders.

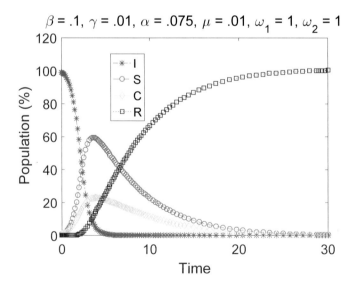

FIGURE 7.20: ISCR: Spreaders dominate.

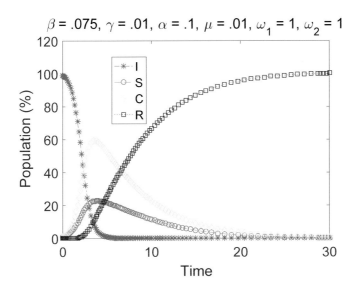

FIGURE 7.21: ISCR: Counter-spreaders dominate.

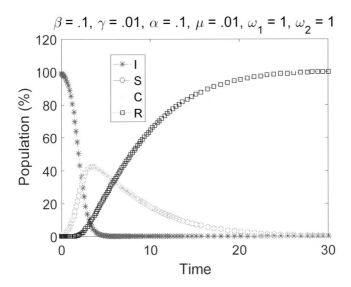

FIGURE 7.22: ISCR: Even mix of spreaders and counter-spreaders.

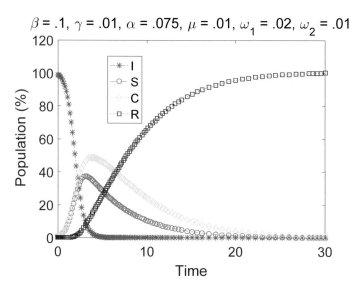

FIGURE 7.23: Spreader twice as receptive to outside influence as a counter-spreader.

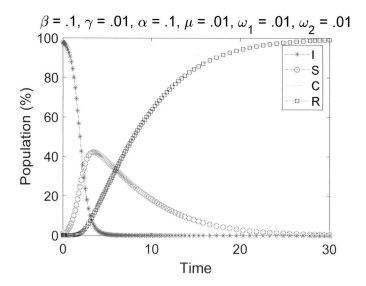

FIGURE 7.24: Heavily-dominant spreader is much more receptive to outside influence compared to a counter-spreader.

(the information in opposition to the main information) holds sway over the initial information being spread. This point occurs once the spreader receptivity to outside views reaches approximately ten times the receptivity of the counter-spreaders to agreeing with the initial information (in this sample case). In other words, being open to the ideas of other social media users (especially those that you disagree with), helps their opposing ideas spread. With a more open-minded communication stream, ideas can flow back and forth more freely and lead to less extreme polarization.

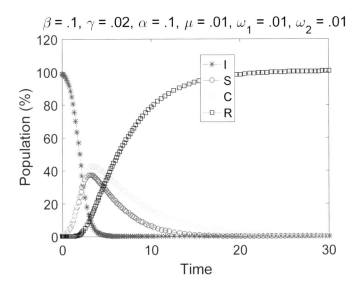

$\beta = .1$, $\gamma = .02$, $\alpha = .1$, $\mu = .01$, $\omega_1 = .01$, $\omega_2 = .01$

FIGURE 7.25: Spreader is stifled twice as strongly as a counter-spreader.

By observing the sample simulations with parameter changes in Figure 7.25 and Figure 7.26, we can see that the spreading and stifling strength changes (with equal receptiveness between opposing ideas) have a very significant effect, which was to be expected. Naturally, a spreading or counter-spreading group with either more interesting information or information that is strongly believed will dominate over one that is boring or untrustworthy.

It should be noted that counter-spreaders are not likely to be generally open-minded to messages that has been demonstrated to be objectively untrue such as in the case of misinformation or disinformation (assuming the source of disproof is well-trusted). As such, the counter-spreader receptivity rate (ω_2) will be low compared to that of the spreaders. This situation is common concerning information that has not been properly researched or fully absorbed, but is nonetheless reacted to and spread over social media. "Fake news" and misinformation concerning scientific research or statistics and confusion over public policy are examples of such a scenario. If a social media user reads a sensationalist "click bait" article headline and reads only the comments (not the actual article), knee-jerk reactions and personal interpretations spread

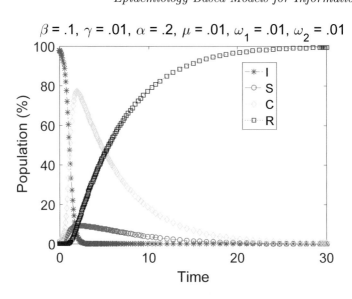

FIGURE 7.26: Counter-spreader is spreading twice as strongly as a spreader.

through the group until eventually the truth is discovered and such debate dwindles.

```
     Code 7.7: Ignorant-Spreader-Counterspreader-Recovered Model
1  clear;
2  to = 0; % initial time
3  tf =30; % final time
4  yo = [98 1 1 0]; % initial conditions of I,S,C,R
       respectively
5  [t y] = ode45('ypISCR',[to tf],yo);
6  plot(t,y(:,1),'-.r*',t,y(:,2),'--mo',t,y(:,3),'-dc
       ',t,y(:,4),':bs')
7  title('\beta = .1, \gamma = .01, \alpha = .1, \mu
       = .01, \omega_1 = .01, \omega_2 = .01', '
       FontWeight','Normal')
8  xlabel('Time')
9  ylabel('Population (%)')
10 legend('I','S','C','R','Location','best')
11 ax = gca;
12 ax.FontSize = 16;
```

```
function ypISCR = ypISCR(t,y)
%This function sets up the dynamics of the system
    to be solved in ODE45
b = .1; %beta = spread rate
g = .01; %gamma = stifle rate
a = .1; %alpha = counter-spread rate
m = .01 ; %mu = counter-stifle rate
k = .2 ; %average number of contacts of each
    individual
w1 = .01; %omega1 = spreader openness to counter-
    spreader
w2 = .01; %omega1 = counter-spreader openness to
    spreader
% Defines the dynamics
ypISCR(1) =-b*k*y(1)*y(2)-a*k*y(1)*y(3); %ignorant
ypISCR(2) = b*k*y(1)*y(2)-w1*k*y(2)*y(3)+w2*k*y(2)
    *y(3)-g*k*y(2)*(y(2)+y(4)); %spreader
ypISCR(3) = a*k*y(1)*y(3)+w1*k*y(2)*y(3)-w2*k*y(2)
    *y(3)-m*k*y(3)*(y(3)+y(4)); %counter
ypISCR(4) = g*k*y(2)*(y(2)+y(4))+m*k*y(3)*(y(3)+y
    (4)); %recovered
ypISCR = [ypISCR(1) ypISCR(2) ypISCR(3) ypISCR(4)
    ]';
end
```

7.9   Hybrid ISCR Model

Often, situations arise where two (or more) primary group "communities" dominate their view and spread of a certain type of information. In the real world, this is a common situation. For example, in the United States, conservatives and liberals often form information communities that oppose one another on political information and interpretation. One country may be a mega-community and oppose a similar community from another country in terms of accepting and spreading news relating to inter-country relations. The ISCR model covers situations where potentially contentious information is being spread within a mostly homogeneous population. However, what it does not do well is describe how information spreads or diffuses between two different, especially polarized groups. It is, therefore, necessary to modify the ISCR model to account for a hybrid case [63].

In the hybrid model, for simplicity, we assume two main polarized communities, $ISCR_1$ and $ISCR_2$, which can be extended to any number of polarized

TABLE 7.6: Parameters for the hybrid ISCR model.

Term	Meaning
$ISCR_x$	ISCR group x
$ISCR_y$	ISCR group y
ω_x	Internal receptivity factor of group x
ω_y	Internal receptivity factor of group y
a_{xy}	Directed information flow factor from group x to y

groups. We assume there are some individuals within each community that have contact with the other community members. Perhaps they are moderates or friends who subscribe to a different political view or a family member from a home county; individuals who are not part of their current information community but act as a link to another community. No matter the reason, some individuals act as diffusion elements of information between the two communities. The influence and connection strength of the overall collection of these individuals is denoted by a pair of directional constants a_{12} and a_{21}, where a_{12} is the strength of information flow from group $ISCR_1$ to group $ISCR_2$ and a_{21} is the strength of information flow from group $ISCR_2$ to group $ISCR_1$, expressed as a percentage of cross-group receptivity. These additional parameters to the previous ISRC model are summarized in Table 7.6. For highly polarized groups, these constants will be tiny fractions of one. For completely isolated communities, they will be equal to zero, reducing the hybrid model to the standard ISCR model discussed previously. Figure 7.27 shows the high-level flow diagram for the proposed hybrid ISCR model. Modifying the previously discussed ISCR model, the dynamics for one group of hybrid ISCR is given as [63]:

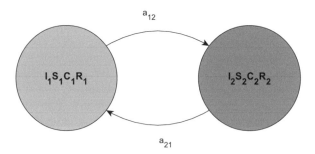

FIGURE 7.27: Flow diagram for ISCR two-community interactions.

$$\dot{i}_1(t) = -\beta_1 i_1(t)s_1(t) - \alpha_1 i_1(t)c_1(t) - a_{21}\beta_2 i_1(t)s_2(t) - a_{21}\alpha_2 i_1(t)c_2(t)$$
$$\dot{s}_1(t) = \beta_1 i_1(t)s_1(t) - \omega_{11}s_1(t)c_1(t) + \omega_{12}s_1(t)c_1(t) - \gamma_1 s_1(t) + a_{21}\beta_2 i_1(t)s_2(t)$$
$$\dot{c}_1(t) = \alpha_1 i_1(t)c_1(t) + \omega_{11}s_1(t)c_1(t) - \omega_{12}s_1(t)c_1(t) - \mu_1 c_1(t) + a_{21}\alpha_2 i_1(t)c_2(t)$$
$$\dot{r}_1(t) = \gamma_1 s_1(t) + \mu_1 c_1(t).$$

$$(7.14)$$

Likewise, for the second ISCR group the following dynamics are given as:

$$\dot{i}_2(t) = -\beta_2 i_2(t)s_2(t) - \alpha_2 i_2(t)c_2(t) - a_{12}\beta_1 i_2(t)s_1(t) - a_{12}\alpha_1 i_2(t)c_1(t)$$
$$\dot{s}_2(t) = \beta_2 i_2(t)s_2(t) - \omega_{21}s_2(t)c_2(t) + \omega_{22}s_2(t)c_2(t) - \gamma_2 s_2(t) + a_{12}\beta_1 i_2(t)s_1(t)$$
$$\dot{c}_2(t) = \alpha_2 i_2(t)c_2(t) + \omega_{21}s_2(t)c_2(t) - \omega_{22}s_2(t)c_2(t) - \mu_2 c_2(t) + a_{12}\alpha_1 i_2(t)c_1(t)$$
$$\dot{r}_2(t) = \gamma_2 s_2(t) + \mu_2 c_2(t).$$

$$(7.15)$$

Following the diffusion terms in the dynamics, the system can be loosely described as individuals from one group spreading their group's information and opinions on a topic to an opposing community via their friends or relatives. The receptiveness of one group to believe and accept the views of another will determine how much the extra-group spreader influences the opposing group.

Consider the two $ISCR$ groups shown in Figure 7.28. The $ISCR_1$ community is predominantly influenced by the spreader information with minor counter-spreader information influence. In contrast, the opposite is true for the $ISCR_2$ community, being influenced mostly by the counter-spreader information (generally the information that contradicts or opposes the spreader information). For example, $ISCR_1$ spreaders might be mainly spreading an unverified rumor about an opposing political party candidate in the $ISCR_2$ community. Some members of $ISCR_1$ believe the initial information is a lie and are counter-spreaders in that community. The $ISCR_2$ community has its own contentious information concerning an opposing candidate from the $ISCR_1$ community with similar dissent levels. Since the a_{12} and a_{21} constants are zero, there is no diffusion with which to cause interaction between the communities.

With an equal amount of cross-interactions between the two communities (Figure 7.29), the influence of counter-spreaders from $ISCR_2$ counteracts the influence of $ISCR_1$ spreaders and vice-versa. This, of course, assumes that in the polarized groups, the information of $ISCR_1$ spreaders is the same message as $ISCR_2$ counter-spreaders. This need not be the case but serves to simplify the system, for example, purposes.

When there is mismatched receptivity between the communities, dramatic shifts can occur in the way popular beliefs are altered via information spread. In Figure 7.30, the $ISCR_1$ community is more receptive to the $ISCR_2$ community and is hence more likely to listen to the messages of the individuals acting as diffusion spreaders and counter-spreaders. In this case, because the counter-spreaders are dominant in the $ISCR_2$ group (and many of these counter-spreaders are being heard), there is a large resurgence of counter-spreaders in $ISCR_1$, while keeping the spreaders of $ISCR_2$ relatively low (as they are not nearly as receptive to outside influence).

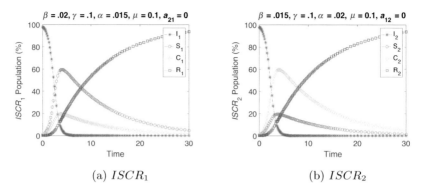

(a) $ISCR_1$ (b) $ISCR_2$

FIGURE 7.28: Sample hybrid ISCR groups: dominant spreader and counter-spreader.

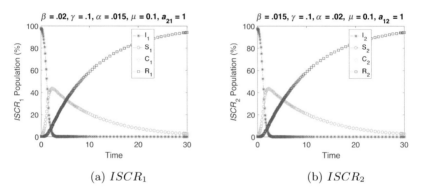

(a) $ISCR_1$ (b) $ISCR_2$

FIGURE 7.29: Sample hybrid ISCR groups: equal bidirectional diffusion.

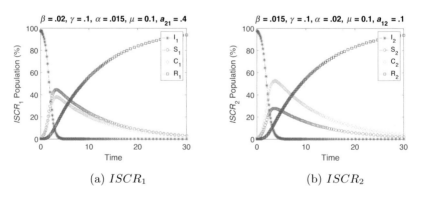

(a) $ISCR_1$ (b) $ISCR_2$

FIGURE 7.30: Sample hybrid ISCR groups: skewed receptivity between the groups.

As skewed receptivity mismatching increases to where one group is about seven times as receptive as the other, we find a "tip over point" where (all else being equal) the opposing community's information beliefs equal that of the primary community (Figure 7.31). After this point, the information beliefs of the opposition group begin to dominate both groups (Figure 7.32). It should be noted that there are significant diminishing returns on the effect of receptivity to opposing polarized communities.

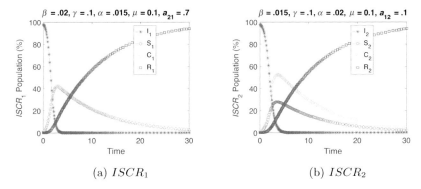

(a) $ISCR_1$ (b) $ISCR_2$

FIGURE 7.31: Sample hybrid ISCR groups: receptivity tipping point.

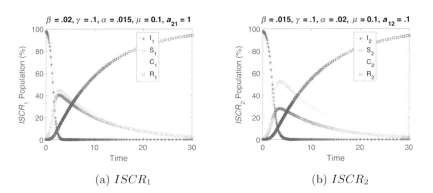

(a) $ISCR_1$ (b) $ISCR_2$

FIGURE 7.32: Sample hybrid ISCR groups: skewed receptivity between the groups.

Clearly, between polarized groups, diffusion plays a significant role in changing the dominant information beliefs of a community. Generally speaking, if two groups are genuinely polarized, outside influences will be felt, but the impact will be minor. A relatively substantial amount of receptiveness to external opposing communities would be required to have lasting change. A situation in which one group is several times more receptive than the other is unlikely. Still, it is of value to objectively observe and model the influence groups have on one another along a spectrum of non-mutual receptivity levels.

Code 7.8: Hybrid ISCR Model

```
clear;
to = 0; %initial time
tf =30; %final time
yo = [98 1 1 0 98 1 1 0]; %initial conditions of
    I1,S1,C1,R1,I2,S2,C2,R2 respectively
[t y] = ode45('hybrid_ypsirc',[to tf],yo);
subplot(2,1,1);
plot(t,y(:,1),'-.r*',t,y(:,2),'--mo',t,y(:,3),'-dc
    ',t,y(:,4),':bs')
title('\beta = .02, \gamma = .1, \alpha = .015, \
    mu = 0.1, a_{21} = 1')
xlabel('Time')
ylabel('ISCR_1 Population (%)')
legend('I_1','S_1','C_1','R_1','Location','best')
ax = gca;
ax.FontSize = 16;
subplot(2,1,2)
plot(t,y(:,5),'-.r*',t,y(:,6),'--mo',t,y(:,7),'-dc
    ',t,y(:,8),':bs')
title('\beta = .015, \gamma = .1, \alpha = .02, m
    = 0.1, a_{12} = .1')
xlabel('Time')
ylabel('ISCR_2 Population (%)')
legend('I_2','S_2','C_2','R_2','Location','best')
ax = gca;
ax.FontSize = 16;
```

```
function hybrid_ypsirc = hybrid_ypsirc(t,y)
%This function sets up the dynamics of the system
    to be solved in ODE45
b1 = .02; %beta1 = spread rate
g1 = .1; %gamma1 = stifle rate
a1 = .015; %alpha1 = counter-spread rate
m1 = .1; %mu1 = counter-stifle rate
w11 = .01; %omega11 = spreader openness to counter
    -spreader
w12 = .01; %omega12 = counter-spreader openness to
    spreader
a21 = 1; %a21 = receptiveness to group 2
b2 = .015; %beta2 = spread rate
g2 = .1; %gamma2 = stifle rate
```

```
12  a2 = .02; %alpha2 = counter-spread rate
13  m2 = .1; %mu2 = counter-stifle rate
14  w21 = .01; %omega21 = spreader openness to counter
        -spreader
15  w22 = .01; %omega22 = counter-spreader openness to
        spreader
16  a12 = .1; %a12 = receptiveness to group 1
17  %SIRC1
18  hybrid_ypsirc(1) =-b1*y(1)*y(2)-a1*y(1)*y(3)-a21*
        b2*y(1)*y(6)-a21*a2*y(1)*y(7); %ignorant 1
19  hybrid_ypsirc(2) = b1*y(1)*y(2)-w11*y(2)*y(3)+w12*
        y(2)*y(3)-g1*y(2)+a21*b2*y(1)*y(6); %spreader
        1
20  hybrid_ypsirc(3) = a1*y(1)*y(3)+w11*y(2)*y(3)-w12*
        y(2)*y(3)-m1*y(3)+a21*a2*y(1)*y(7); %counter 1
21  hybrid_ypsirc(4) = g1*y(2)+m1*y(3); %recovered 1
22  %SIRC2
23  hybrid_ypsirc(5) =-b2*y(5)*y(6)-a2*y(5)*y(7)-a12*
        b1*y(5)*y(2)-a12*a1*y(5)*y(3); %ignorant 2
24  hybrid_ypsirc(6) = b2*y(5)*y(6)-w21*y(6)*y(7)+w22*
        y(6)*y(7)-g2*y(6)+a12*b1*y(5)*y(2); %spreader
        2
25  hybrid_ypsirc(7) = a2*y(5)*y(7)+w21*y(6)*y(7)-w22*
        y(6)*y(7)-m2*y(7)+a12*a1*y(5)*y(3); %counter 2
26  hybrid_ypsirc(8) = g2*y(6)+m2*y(7); %recovered 2
27  hybrid_ypsirc = [hybrid_ypsirc(1) hybrid_ypsirc(2)
        hybrid_ypsirc(3) hybrid_ypsirc(4)
        hybrid_ypsirc(5) hybrid_ypsirc(6)
        hybrid_ypsirc(7) hybrid_ypsirc(8)]';
28  end
```

7.10   ISSRR Model for Contentious Information

The ISCR model presented in this text depicts two dominant groups in which class members with opposite viewpoints wish to persuade the other to their beliefs concerning contentious information, it does not model a scenario in which there exist more than one recovered state (over the contentious information). For example, suppose there is a community vote over Measure Z to add funding to a social program at the expense of increasing taxes. Is the social program worth the tax increase or can the money be spent on something more valuable to the community? After some back-and-forth debate, community members choose which side of the debate they ultimately agree with, stop debating, and vote on the measure. Since the previous ISCR models do not address multiple end-points of contentious information spread, a special model must be developed to examine this form of spread in a social media network.

Generally speaking, we consider the basic social media focused ISR modeling of a polarized social network group presented with contentious information, leading to two final recovered beliefs. The two recovered groups represent individuals who eventually make up their minds regarding the story and will ultimately lose excitement in debating it further (at least pending any pertinent new information).

To begin, the social network group discovers the latest news (learning about Measure Z from the previous example) and people form an initial assessment. Several people will opt to begin spreading their opinions among their friends and fellow social network members via tweets, shares, and memes and two spreader groups will emerge. At this point, each spreader group will use shares, posts, and tweets to convince others to come to their "side" of the contentious information (campaigning for or against Measure Z in our example). Those people who change their minds might loosely campaign as spreaders for their new stance. Once internal debates have concluded, each community member finally settles on a favored stance between the two recovered states (and vote "yes" or "no" on Measure Z in our example). We describe this community interaction scenario through an Ignorant-Spreader-Spreader-Recovered-Recovered (ISSRR) model, which is illustrated in Figure 7.33.

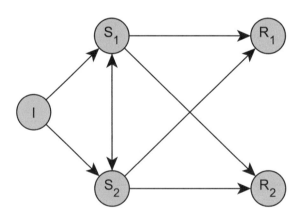

FIGURE 7.33: Class interactions for the ISSRR model.

Observe that for this model, after individuals become informed of the topic and divide into their respective spreader classes, they will still interact (with different degrees of balance) with opposing spreaders and ultimately settle into a final recovered state, favoring one of the contentious sides. While the ongoing trend of members of the ignorant class transforming into spreaders remains the same as with previously examined models, spreader-spreader interactions must be considered in a new light, taking into account the tendency to (after some debate) settle into one of the two recovered belief stances, based on that spreader-spreader interaction.

Simply put, this new ISSRR model tailored for online social media focuses on which side has the greater capability to convince the other. This ability to compel can be based on a number of factors: demonstrable facts uncovered that resolve the contention, assumed authority figures from the spreader group, and other factors. These factors are weighed against the tendency of a community to maintain their original beliefs, which will eventually decide the overall result between the competing classes of message circulation and establish a prevailing position.

The proposed ISSRR model takes the form of the following set of differential equations:

$$\begin{cases} \dot{i}(t) = -\beta_1 i(t)s_1(t) - \beta_2 i(t)s_2(t) \\ \dot{s}_1(t) = \beta_1 i(t)s_1(t) + (d_{12} - d_{21})s_1(t)s_2(t) - (\gamma_{11} + \gamma_{12})s_1^2(t) \\ \dot{s}_2(t) = \beta_2 i(t)s_2(t) + (d_{21} - d_{12})s_1(t)s_2(t) - (\gamma_{22} + \gamma_{21})s_2^2(t) \\ \dot{r}_1(t) = \gamma_{11}s_1^2(t) + \gamma_{21}s_2^2(t) \\ \dot{r}_2(t) = \gamma_{22}s_2^2(t) + \gamma_{12}s_1^2(t), \end{cases} \quad (7.16)$$

where the relative diffusion influence and strength of one spreading community over the other is d. Notice that like other social media interaction models presented, it is assumed that the recovered class does not actively engage in further pairwise interactions. As both spreading groups debate with each other and spread to the ignorant class, the recovered group will, at best, "lurk" or spectate without contributing to the conversation because they have exhausted their active interest in discussions. Like the other models, modified for modern social media applications, this model deviates from conventional rumor spread models.

Two example situations are simulated using the dynamics presented for the ISSRR model. For the first example, one of the groups is noticeably better at spreading their message than the other over online social media, as shown in Figure 7.34. Several reasons can be postulated as to why there is this disparity. The dominant group is more comfortable with internet communication or more open to share opinions over the internet for cultural or generation-related reasons. For the sake of comparison, both spreading groups are assumped to be equally likely to lose interest in the discussion their minds have been made up concerning the debate and are also both assumed to equally favor their starting group position and disfavor the opposing group's opinion. Finally, both groups have a balanced influence over each other within their social media network. The assumptions presented allow examination of the ability to spread more effectively in isolation. Observe that a higher spreading effectiveness leads to one group's significant influence over the final total community's recovered opinion distribution.

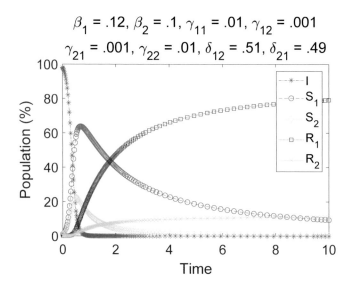

$$\beta_1 = .12,\ \beta_2 = .1,\ \gamma_{11} = .01,\ \gamma_{12} = .001$$
$$\gamma_{21} = .001,\ \gamma_{22} = .01,\ \delta_{12} = .51,\ \delta_{21} = .49$$

FIGURE 7.34: ISSRR scenario: strong Group 1 spreading.

For the second example, displayed in Figure 7.35, both spreader groups show equal effectiveness at transmitting their position and equally likely to lose interest with either opinion, once they have decided on a topic position. In contrast with the first example, here, the second community subgroup has stronger influence over the opposing spreader subgroup. This scenario quickly results in much higher numbers of community members settling on the second group's position on the informationn.

Several real-world examples demonstrate the practicality of presenting a novel ISSRR model. A community must sometimes face social and cultural issues that have no objective right or wrong answer. Political campaigns between competing measures or candidates that represent a community ideology split will incite debate and lead to subjective decisions. Additionally, fake news and misinformation concepts can be explored using this model, assuming the competing news stories are championed by highly polarized primary groups, where final-state recovered individuals conclude that the news is either legitimate information or misinformation.

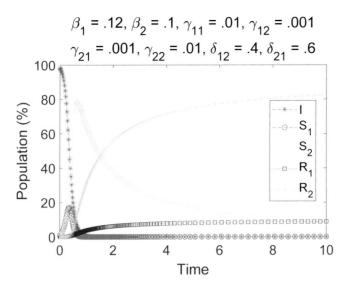

FIGURE 7.35: ISSRR scenario: minor influence of Group 2 over Group 1.

```
                    Code 7.9: ISSRR Model
1  clear;
2  to = 0; %initial time
3  tf =10; %final time
4  yo = [98 1 1 0 0]; %initial conditions of I,S1,S2,
       R1,R2 respectively
5  [t y] = ode45('ypISSRR',[to tf],yo);
6  plot(t,y(:,1),'-.r*',t,y(:,2),'--mo',t,y(:,3),'-dc
       ',t,y(:,4),':bs',t,y(:,5),'-gx')
7  title({'\beta_1 = .12, \beta_2 = .1, \gamma_{11} =
       .01, \gamma_{12} = .001', '\gamma_{21} =
       .001, \gamma_{22} = .01, \delta_{12} = .51, \
       delta_{21} = .49'}, 'fontweight', 'normal')
8  xlabel('Time')
9  ylabel('Population (%)')
10 legend('I','S_1','S_2','R_1','R_2','Location','
       best')
11 ax = gca;
12 ax.FontSize = 16;
```

```
1  function ypISSRR = ypISSRR(t,y)
2  %given the input parameters of an SIRC model, this
       function sets up the dynamics of the
3  %system to be solved in ODE45
4  b1 = .12;  % beta_1 spread constant
5  b2 = .1; % beta_2 spread constant
6  g11 = .01; % gamma stifle constant from S1 to R1
7  g12 = .001; % gamma stifle constant from S1 to R2
8  g21 = .001; % gamma stifle constant from S2 to R1
9  g22 = .01; % gamma stifle constant from S2 to R2
10 d12 = .51; % influence of group 1 on group 2
11 d21 = .49; % influence of group 2 on group 1
12
13 ypISSRR(1) =-b1*y(1)*y(2)-b2*y(1)*y(3); %ignorant
14 ypISSRR(2) = b1*y(1)*y(2)+(d12-d21)*y(2)*y(3)-(g11
       +g12)*y(2)^2; %spreader 1
15 ypISSRR(3) = b2*y(1)*y(3)+(d21-d12)*y(2)*y(3)-(g22
       +g21)*y(3)^2; %spreader 2
16 ypISSRR(4) = g11*y(2)^2+g21*y(3)^2; %recovered 1
17 ypISSRR(5) = g22*y(3)^2+g12*y(2)^2; %recovered 2
18 ypISSRR = [ypISSRR(1) ypISSRR(2) ypISSRR(3)
       ypISSRR(4) ypISSRR(5)]';
19
20 end
```

7.11 Exercises

1. Explain the differences between the IS and the ISI models. When is each appropriate to use? Give examples.

2. There are already well-established models used to describe person-to-person rumor propagation within a population. How is the spread of information different in an online environment? How can this be accounted for mathematically?

3. Compare the spread of information using the ISR model for the three cases when reproduction number R_0 is: (a) 0.5, (b) 2, and (c) 5. Report the percentage of the ignorant class that receives the information in each case. You may assume suitable parameters and use the code provided in Section 7.5 for this exercise.

4. Develop a `MATLAB` code for simulating the ISR model with decay, as described in Section 7.7.1. You may assume suitable parameters and use the provided code for this exercise.

5. Develop a `MATLAB` code for simulating the ISCR model as described in Section 7.8. You may assume suitable parameters such that there is an even mix of spreaders and counter-spreaders, spreaders are three times as receptive to outside influence as counter-spreaders, and counter-spreaders are stifling twice as strongly as spreaders. You may use the sample code provided in Section 7.8 for this exercise.

6. Compare and contrast the hybrid-ISCR model with the ISSRR model. Identify a real-life example for each one of these and the reason why one model would make more sense to apply than the other, given your example.

7. Read the case study: "2014 South Napa Earthquake" from the reference listed below and debate which information spread model might be most suitable in this case. If none of the models are found appropriate, list the modifications required in existing models. Some highlights from the case study are as follows: *"On August 24, 2014, at 3:20 a.m., a 6.0-magnitude earthquake struck the area of American Canyon and Napa, California. The earthquake shook awake many residents in the Bay Area and provoked a nearly instant social media response, particularly on Twitter. As information about the earthquake became available online, the hashtags # NapaQuake*

and # NapaEQ were broadly used by people in the affected area and those responding to the earthquake. As hashtags become popular on Twitter, spammers and troll target those hashtags in an effort to have a broader audience for their unrelated message. In the case of # NapaQuake, a particularly disturbing "hashtag hijacking" took place. For much of the first days of the earthquake response, a significant portion of the tweets on the most popular response hashtags contained graphic pictures of dead bodies from unrelated events. The main subject of the hijacking tweets were accusations of U.S. military misconduct with images of people being tortured or horribly mangled bodies being included as evidence. This was shocking content for social media monitors who were used to dealing with more standard disaster response tweets, not inflammatory and graphic material. "

8. Read the case study: "2016 Louisiana Floods" from the reference listed below and debate which information spread model might be most suitable for application in this case. If you decide no models are appropriate, list the modifications required to existing models. Some highlights from the case study are as follows: "*ViaLink Louisiana, a 2-1-1 provider, found itself overwhelmed with calls following the March 2016 floods in Louisiana. ViaLink noticed multiple inaccurate Facebook messages and posts that went viral and contributed to the number of calls. After FEMA declared a disaster, the calls kept coming. In addition to the continuing Facebook messages and posts, FEMA was also giving out the incorrect information and referring people to 2-1-1 for claim assistance (the information was later corrected). In a similar situation, during the response to Louisiana's summer floods in 2016, the American Red Cross was confronted with multiple rumors and misinformation on social media related to its shelter policies and food distribution. False claims spread especially rapidly through new video tools, such as Facebook Live, and threatened to erode the public's trust and support, as well as eclipse the personalized care and outreach that the organization was providing through social media.* "

Reference: "Countering False Information on Social Media in Disasters and Emergencies", a report by *Social Media Working Group for Emergency Services and Disaster Management, Homeland Security*. Published March 2018. $https$: $//www.dhs.gov/sites/default/files/publications/SMWG_{Countering-} False-Info-Social-Media-Disasters-Emergencies_{M}ar2018-$ 508.pdf

8

Stochastic Modeling of Information Spread

> *We define the art of conjecture, or stochastic art, as the art of evaluating as exactly as possible the probabilities of things, so that in our judgments and actions we can always base ourselves on what has been found to be the best, the most appropriate, the most certain.*

Jacob Bernoulli, *Ars Conjectandi*, 1713

In the previous chapter, epidemiology-based deterministic models were presented and examined. At each point in time, these models assumed deterministic state values to represent the system. While considering a deterministic value for a state is very convenient in simplifying a complicated system, real-world systems for information spread (and most systems in general) are naturally probabilistic. In probabilistic systems, simulations and models must consider a range of variable values under a reasonable probability distribution. Models that account for the probabilistic nature of these variables are known as stochastic models and are widely used in mathematics, physics, natural sciences, manufacturing, medicine, and more.

In this chapter, the concept of stochastic processes and models are explored, beginning with a visual and textual explanation of Brownian motion particles. A basic function that should be familiar to all readers is presented deterministically. For comparison, the concept of a stochastic realization is presented for the same basic function. Building upon the idea of stochasticity in systems, applications for social media systems are considered. Specifically, ISI and ISR model dynamics for social media information spread are derived, explained, and simulated. Sample MATLAB code is provided to simulate these and other stochastic online social media information spread processes. With a basic understanding of stochastic processes, similar methods can be applied to other models throughout the text.

8.1   Brownian Motion

Before examining stochastic models, it is first helpful to understand stochastic processes in general. One of the most basic stochastic processes is known as Brownian motion. The process was first described in 1827 by botanist Robert Brown while observing pollen immersed in water via a microscope. It is the random motion of particles suspended in a liquid or gas. As these particles move, they collide with the quickly moving molecules within the fluid and change direction. Since the force and direction of the particles are always changing, and the suspended particles collide at different angles, the result is seemingly random particle motion and paths [65]. Refer to Figure 8.1 for a visual representation of Brownian motion in a closed system. Figure 8.2 shows a simulation of two Brownian motion particles in three-dimensional space.

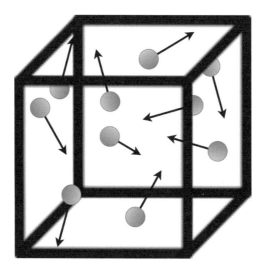

FIGURE 8.1: Brownian motion of large particles.

Deterministic models cannot solve the physical interactions in Brownian motion because such models cannot account for each molecule and particle position, path, and interaction within the process. As such, only probabilistic models (such as stochastic process models) can effectively describe Brownian motion and other such processes.

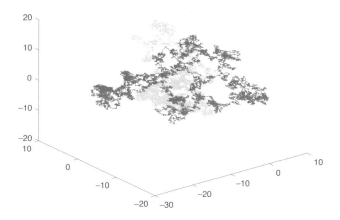

FIGURE 8.2: Simulation of Brownian motion for two particles.

8.2 Deterministic and Stochastic Realizations of Processes

Consider a simple sinusoidal function, $y = sin(t)$. From a purely deterministic mathematical perspective, each value of time will lead to a predictable corresponding value. As time progresses, the set of time values results in the formation of a clear and consistent sin wave path. It is known as the "deterministic path" and can be seen as the dashed line in Figure 8.3.

Now let's examine a real-world sinusoidal such as an ocean wave. While such waves can be modeled deterministically, the reality is far messier. Experimentally, ocean waves are *not* perfect sine waves. Not only will the wave's crests and troughs be inconsistent, but also the curve itself. Small subsections of the wave will crash into other small parts, and splashes will result, adding further chaos to any notion of an ideal wave shape and making it's traced path probabilistic. The reason for this stochastic behavior is external and unpredictable influences. In the case of the ocean wave, these influences might be air currents hitting the water's surface, marine life, nearby cruise ships, or any number of other elements. Because each of these external elements cannot be precisely known or integrated into a deterministic model, it must be treated as a stochastic process with several probability-based paths or realizations. Any such actual path taken by one of these stochastic processes is known as a "stochastic realization". A sample stochastic realization of the sine function example is shown overlapping the deterministic solution in Figure 8.3.

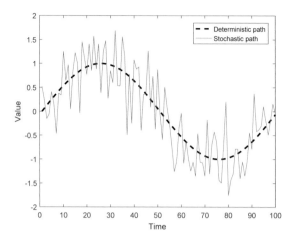

FIGURE 8.3: Stochastic realization of a sinusoidal signal.

8.3   Stochastic Modeling Considerations for Social Media Systems

In social science and social theory, the stochasticity of events and interactions involving people must also be considered. Here, the unconscious processes and decisions of human behavior contribute to the system's probabilistic and external influences. Each event or interaction will have its own unpredictable outcome due to the number of human-based variables involved.

Although stochastic models significantly add to the complexity of a system model, both conceptually and mathematically, they also grant important insight into the bounds of the likely behavior of a system. It is neither intuitive nor reasonable to assume that an information spread model will completely predict (down to each person) the time progression and overall diffusion of any given message as it spreads over social media. Why? Because human beings by nature are unpredictable and probabilistic. Given this, upon modeling many human-critical systems, a stochastic-minded approach to modeling is wise. As an example, proposed tax rates for vehicles based on miles traveled should be modeled using stochastic differential equations because all people do not have identical driving habits or vehicles [66].

In the remaining sections of this chapter, two stochastic information spread models are presented: a traditional ISI information spread model and an ISR information spread model, tailored specifically to social media information [63]. Similar to the other information spread models presented this text, the models here are adapted from mathematical epidemiology models [67] with particular attention paid to variable naming conventions and differences in state transitions due to people's behavior over social media. A trivially small birth rate and death rate is assumed due to the rapid spread of information over social media. Similar fundamental ISI and ISR model principles can be applied to any other epidemic-based information spread models to obtain their corresponding stochastic model.

8.4 Stochastic ISI Information Model

Let's take a moment and review the Ignorant-Spreader-Ignorant interactions of the standard ISI model. Through the dynamics of the ISI model, ignorant class members learn and gain interest in a particular topic. They then pass the information to other members of the community until the information grows old or becomes dull. Over time, several members of the spreader group return to the ignorant state until a the topic or associated group of topics has new and exciting information available to revitalize an urge to distribute it. The stochastic ISI model does not diverge conceptually from this core process. However, stochastic systems must be approached in a different way in light of their probabilistic nature.

Typically, $N = 1$ for simplicity is a simulation of population percentage. Here, we expand the ISI deterministic system and generalize N such that $N = I + S$ as the total community population being observed. Further simplifying for discussion, group classes can be represented without time t unless specified, though the classes still remain a function of time. The dynamics of the IS system here are defined as:

$$\dot{i} = -\frac{\beta}{N}is + \gamma s$$
$$\dot{s} = \frac{\beta}{N}is - \gamma s .$$

(8.1)

Now, we shall include the following Itô SDE:

$$\frac{dS}{dt} = \mu(S) + \sigma(S)\frac{dW}{dt}.$$

(8.2)

In this equation, W is the stochastic Wiener process. The system model converges to the Itô SDE if if the Euler method is used and specific growth and smoothness conditions are met. This method is the same as that used in stochastic epidemic models [67]. In this ISI model, $\mu(S) = b(S) - d(S)$ is the deterministic growth and decay of the information (the same as before), and $\sigma(S) = \sqrt{b(S) + d(S)}$ is the stochastic component of the system, where

$$b(S) = \frac{\beta}{N}S(N - S)$$
$$d(S) = \gamma S.$$

By conveying the dynamics in terms of the spreader class and making the previously identified substitutions into the Itô SDE, the stochastic ISI information spread model can be written in an SDE form as follows:

$$\frac{dS}{dt} = \frac{\beta}{N}S(N-S) - \gamma S + \sqrt{\frac{\beta}{N}S(N-S) + \gamma S}\frac{dW}{dt}. \qquad (8.3)$$

We now plot the stochastic ISI dynamics in Figure 8.4 with a set of stochastic sample realizations. At this point we can see that the example simulations closely mirror the anticipated deterministic ISI dynamics when considered in their entirety. Individually, however, they deviate from one another.

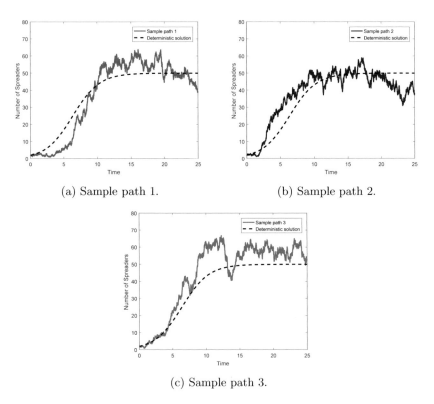

(a) Sample path 1.   (b) Sample path 2.

(c) Sample path 3.

FIGURE 8.4: Stochastic realizations of an ISI network.

8.5 Stochastic ISR Information Modeling and Social Media

Assumptions and processes common to other epidemic models discussed can be utilized for the ISR information spread model, as was done with the stochastic ISI model from the previous section. Again, a fairly normal distribution is assumed for the random variables. Here, we explore two different cases of the ISR information spread model: the conventional (person-to-person) ISR rumor model and the proposed modified model applied toward social media [63].

Like with the standard ISR model, the ignorant class here discovers news stories or messages from both standard media and online social avenues, including television, newspaper articles, in-person word-of-mouth discussion, and social media sources, and then spreads the information to others within the community. Again, the rates of births and deaths associated with similar epidemiology-based models are imperceptible in this context and are omitted due to the rapid speed of agent contact.

If $\Delta X(t) = (\Delta I, \Delta S)^T$, its expectation can be expressed as follows:

$$E(\Delta X(t)) = \begin{bmatrix} -\frac{\beta}{N}IS \\ \frac{\beta}{N}IS - \gamma S[S+R] \end{bmatrix} \Delta t.$$

Next, the expectation can be used to determine the covariance matrix:

$$V(\Delta X(t)) = E(\Delta X(t)[\Delta X(t)]^T) - E(\Delta X(t))E(\Delta X(t))^T$$
$$V(\Delta X(t)) \approx E(\Delta X(t)[\Delta X(t)]^T)$$
$$V(\Delta X(t)) = \begin{bmatrix} \frac{\beta}{N}IS & -\frac{\beta}{N}IS \\ -\frac{\beta}{N}IS & \frac{\beta}{N}IS + \gamma S \end{bmatrix} \Delta t.$$

The covariance matrix is positive definite. It is also symmetric with square root $B\sqrt{\Delta t} = \sqrt{V}$. Further simplifying, the random vector $X(t + \Delta t)$ can be approximated as follows:

$$X(t + \Delta t) = X(t) + \Delta X(t) \approx X(t) + E(\Delta X(t)) + \sqrt{V(\Delta X(t))}. \quad (8.4)$$

This is an Euler approximation to a system of Itô standard differential equations and assuming reasonably smooth coefficients. The solution of $X(t)$ converges to

$$\frac{dI}{dt} = -\frac{\beta}{N}IS + B_{11}\frac{dW_1}{dt} + B_{12}\frac{dW_2}{dt}$$
$$\frac{dS}{dt} = \frac{\beta}{N}IS - \gamma S + B_{21}\frac{dW_1}{dt} + B_{22}\frac{dW_2}{dt}, \quad (8.5)$$

where W_1 and W_2 are independent stochastic Wiener processes and B_{ij} is the environment-based fluctuation intensity. Here, the network profile and population are the main influences [67].

Code 8.1: Stochastic ISI Model

```
%Stochastic Differential Equation
%ISI Epidemic Model
%Three Sample Paths and the Deterministic Solution
clear
beta=1; % spreading rate
b=0.25;
gam=0.25; % forgetting rate
N=100; % population size
init=2;
dt=0.01; % time step
time=25;
sim=3;
for k=1:sim
    clear i %, t
    j=1;
    i(j)=init;
    t(j)=dt;
    while i(j)>0 & t(j)<25
        mu=beta*i(j)*(N-i(j))/N-(b+gam)*i(j);
        sigma=sqrt(beta*i(j)*(N-i(j))/N+(b+gam)*i(
            j));
        rn=randn; % standard normal random number
        i(j+1)=i(j)+mu*dt+sigma*sqrt(dt)*rn;
        t(j+1)=t(j)+dt;
        j=j+1;
    end
    plot(t,i,'r-','Linewidth',2);
    hold on
end
%Euler's method applied to the deterministic ISI
    model.
y(1)=init;
for k=1:time/dt
    y(k+1)=y(k)+dt*(beta*(N-y(k))*y(k)/N-(b+gam)*y
        (k));
end

%Plotting the results
plot([0:dt:time],y,'k--','Linewidth',2);
axis([0,time,0,80]);
xlabel('Time');
ylabel('Number of Spreaders');
legend('Sample path 1','Sample path 2','Sample
    path 3','Deterministic solution');
hold off
```

8.6   Exercises

1. How is a stochastic model different than a deterministic model?

2. List at least three stochastic information spreading systems. Explain and defend your choices.

3. Summarize the concept of Brownian motion. How does it relate to social media-based stochastic systems?

4. Explain the concept of a stochastic realization.

5. Develop code using MATLAB, Python, or any other suitable software to simulate a deterministic function of your choice along with a sample stochastic realization of the same function. The plot should have both the deterministic and stochastic paths, similar to Figure 8.3. *Hint:* use the **randn()** function to simulate randomness in your process.

6. Develop a MATLAB code for simulating a stochastic ISR information spread model as described in Section 8.4. You may assume suitable parameters and use the stochastic ISI epidemic methods for this exercise.

9

Social Marketing-Based Models for Information Spread

In previous chapters, we considered mostly epidemic-based models (or models derived from them) to describe social media information spread. However, other well-studied and similarly practical models can also be used as the starting point of modeling information spread from another angle. In this chapter, we examine and discuss well-established advertising and marketing models. Building upon these models, special considerations are discussed and integrated into elements of these models to accommodate contemporary social media networks in the context of advertising and general information spread over these mediums. Finally, a new modeling framework is proposed and simulated from the perspective of social media marketing, which will be tested in the next chapter using case studies.

One familiar example of typical advertising or marketing scenario is the case of a marketing agent sending out regular commercial broadcasts to people to encourage them to purchase a product or generally subscribe to the advertisement's suggestion. We see these types of advertisements often for pharmaceutical products, the latest smartphone, television shows on the same network, or political campaigns. Over the years, several effective models have been presented and utilized to capture marketing dynamics. After all, a key goal of dispersing a message is to do so efficiently and within the bounds of time and resources, making quantitative models necessary. These messages are often actively spread via constant commercial and billboard ads. They can also be passively spread, where an informative message is simply relayed once and contains no long-term agenda (as with a public service announcement). However, with the advent of social media, a new phenomenon has arisen: people converting themselves into marketing agents after viewing these advertisements. More than any other modern platform, marketing on social media has massive potential to foster product loyalty and brand recognition. Additionally, it presents new ways to acquire new customers or convert them to patronize a competing business [68].

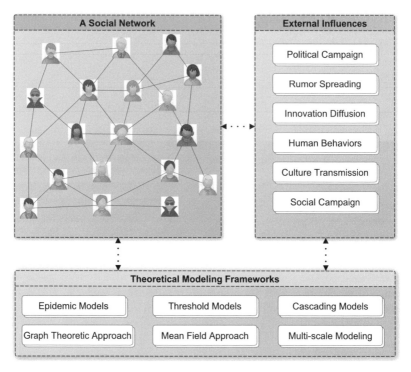

FIGURE 9.1: Social networks, external influences, and theoretical modeling frameworks. Courtesy of [69].

At a high level, how does an organization sell to someone? Typically, individuals or groups (like marketing agencies) are enlisted as message spreaders who broadcast that information in a variety of formats including billboards, social media posts, and television advertisements. Promotion over social media is particularly interesting (and is the topic of this chapter) due to the possibility the advertisement can start "trending" or gaining traction within a group. Questing for special quality often results in billions of dollars in world-wide spending [70]. For our generalized purposes in this text, we consider not only traditional business advertisements but also include actions of individuals that seek to inform and hopefully sway other people to a product, cause, or viewpoint. Without advertisements, any single social campaign or product would be unlikely to sufficiently trend throughout the public consciousness to reach a meaningful percentage of the general population.

Additionally, we must consider what transpires when a social media event or advertisement campaign concludes? Does the message entirely vanish from public awareness? Of course not. With no deliberate publicity, particularly in an online social media environment, talk of the event or social cause and related socio-political issues will continue for a time. Usually, an event, product, or social cause will trigger intentional internet social intercommunication (such as through forums, blogs, memes, or videos). Naturally, this discussion peaks at a certain point and then gradually decays. These communications may persist indefinitely, but will reach a state of relative inaction where only a few individuals discuss the topic or cause until something incites a renewal of interest.

Consider the case of a new gadget, vacation destination, trend, or event. In many real-life instances, these products and activities take on a life of their own as a *"social craze"*. If you see friends and acquaintances pushing the value of these things on social media sites, you might be more inclined to try them out yourself and spread the word to other members of your social network. Traditional marketing models do not handle the impact of social media on marketing as effectively as they do conventional marketing environments, and newer approaches are desirable.

Figure 4.7, which shows the interconnections between social networks, their external influences, and theoretical frameworks useful in modeling the interactions between the two. Epidemic and rumor diffusion models are well investigated and work well for a mixed-population in conventional communication of person-to-person spreading (or pairwise interactions). Special attention, however, should be given when it involves contemporary social media networks.

As such, this chapter begins with a discussion of three popular and fundamental models from marketing: the Vidale-Wolfe model, the Bass model, and the Sethi model. Furthermore, a novel *event-triggered social media chatter model* is examined from a social marketing perspective [69].

9.1 Vidale-Wolfe Model

We begin by examining the popular and long-established Vidale-Wolfe model. The Vidale-Wolfe model is one of the first advertising models for continuous-time systems and is given by

$$\frac{dS(t)}{dt} = \beta u(t)[M(t) - S(t)] - \delta S(t), \tag{9.1}$$

where S is the sales for a brand and M is the size of the market. Parameter δ is the rate of decay of brand sale given no active advertising [71]. Stated simply, when advertisements target new markets, growth occurs. In the absence of advertising, growth decreases. The Vidale-Wolfe model variables and parameters are summarized in Table 9.1.

TABLE 9.1: Vidale-Wolfe model variables and parameters

Term	Meaning
$S(t)$	Sales at time t
$M(t)$	Market size at time t
β	Advertising constant
δ	Rate of brand sale decay
$u(t)$	Control action at time t

9.2 Bass Model

As one of the most cited works in management science, the Bass model captures the interaction between buyers and an untapped market via traditional person-to-person information spread. This model describes the diffusion of new technology, innovations, and products [71]. Bass's differential equation model is specified as:

$$\frac{dN(t))}{dt} = (p + \frac{q}{M}N(t))(M - N(t)), \tag{9.2}$$

where N is the time-dependent cumulative buyers, p is the innovation coefficient, and q is the imitation coefficient. Notice that the $(M - N(t))$ term is reminiscent of the buyer-market interaction in the Vidale-Wolfe model.

The Bass model has since been generalized to account for advertising and price inputs with the addition of $F(t)$ on the right-hand side of the differential equation, as follows:

$$\frac{dN(t))}{dt} = (p + \frac{q}{M}N(t))(M - N(t))F(t), \tag{9.3}$$

where

$$F(t) = 1 - \alpha\{[\frac{\dot{p}(t)}{p(t)}] + \beta[\frac{\dot{a}(t)}{a(t)}]\}, \tag{9.4}$$

and α and β are the sensitivity parameters for price and advertising price, respectively. The Bass model variables and parameters are summarized in Table 9.2.

9.3 Sethi Model

The Sethi advertising model expands upon the fundamental principles of the Vidale-Wolfe model. The dynamics of the Sethi advertising model take the

TABLE 9.2: Bass model variables and parameters

Term	Meaning
$N(t)$	Cumulative buyers at time t
M	Market size
p	Coefficient of innovation
q	Coefficient of imitation
$F(t)$	Cumulative installed base at time t
α	Price sensitivity parameter
β	Advertising sensitivity parameter

form of a stochastic differential equation as

$$dX_t = (rU_t\sqrt{1 - X_t} - \delta X_t)dt + \sigma(X_t)dz_t, \qquad (9.5)$$

where X_t is the market share at time t, U_t is the advertising rate at time t, r is the coefficient of advertising effectiveness, δ is the decay constant, $\sigma(X_t)$ is the diffusion coefficient, and z_t is the stochastic Wiener process (standard Brownian motion).

We can expand Equation 9.5 through use of the Sorger approximation [72]. This says $\sqrt{1 - X_t} \approx (1 - X_t) + (1 - X_t)X_t$, where the first term is the effect of the untapped market and the second term is effect of word-of-mouth interactions. This leads to an expanded representation of the Sethi model where the stochastic term is neglected for the sake of simplicity:

$$dX_t = (rU_t(1 - X_t) + rU_t(1 - X_t)X_t - \delta X_t)dt. \qquad (9.6)$$

Let's analyze and explain Equation 9.6 in further detail. Beginning with the first term, $rU_t(1 - X_t)$, advertising control action U is acting upon individuals who are not aware of the advertisement, $1 - X_t$, along with a constant parameter r to denote the effectiveness of that same advertisement. Moving on to second term, $rU_t(1 - X_t)X_t$, the identical advertising control action and constant parameter are influencing the social interaction occurring between those oblivious to the advertisement campaign and those already familiar with it, $(1 - X_t)X_t$. The last term, $\delta X_t)$, is a "forgetting factor" representing a loss of interest as time passes according to the parameter δ. In summary, Sethi's advertising model demonstrates that marketing affects those ignorant of an item directly and has an additional impact on those unaware of the item through social interactions. This is paired alongside a forgetting factor as individuals become bored or uninterested in the item. The Sethi model variables and parameters are summarized in Table 9.3.

TABLE 9.3: Sethi model variables and parameters

Term	Meaning
X_t	Market share at time t
U_t	Rate of advertising at time t
r	Advertising effectiveness coefficient
δ	Decay constant
$\sigma(X_t)$	Diffusion coefficient
z_t	Wiener process

9.4 Event-triggered Social Media Chatter Model

None of the contemporary marketing models appear entirely able to adapt to the recent phenomena of online social media. For instance, consider a scenario where advertisements or promotions become self-sustaining "crazes" or "talking points" such as a social media "challenge", a popular meme, or widely accepted misinformation. Naturally, the best place to start in such an endeavor is to examine existing models, see where they succeed, and fall short of the modeling needs and work toward the development of a new model proposal.

Conceptually, the marketing models presented here can be generalized to account for any form of information spread. After all, advertising and marketing is essentially targeted spread of awareness of a service or product. We can apply the Vidale-Wolfe model toward social media marketing by making some important modifications. First, by setting $M = 1$ the "market" will be normalized and represent the total size of the population (the people within a certain social media community). Next the spreader variable S from the epidemic-based models is set to be a generalized term, X_t for a specific social media topic being spread. With these modifications we can see that

$$dX_t = \beta u(t)[1 - X_t] - \delta X_t. \tag{9.7}$$

The normalized Vidale-Wolf model in Equation 9.7 can now be compared to the Sethi model in Equation 9.6. Although the Sethi model has a social interaction term and a marketing term, it is ultimately inappropriate to fully describe an online social media advertising campaign, despite being established as an effective advertising model. Here, the goal is to capture a self-sustaining social chatter, but the term $rU_t(1 - X_t)X_t$ will vanish as soon as an event or active advertisement terminates. In an online social media community, people keep discussing and spreading a product's information, a social movement, or an event that has been sufficiently advertised, even if it doesn't reach a "social craze", "viral", or "trending" status.

Continuing progression using the normalized Vidale-Wolfe model, the constant β can be broken apart into two components. We shall define β_1 as the social marketing campaign constant and β_2 as the "social interaction" constant. The effectiveness of social marketing is affected by dynamic resource spending and promotion over the network to convince people to purchase a product, uphold a social or political movement, or join in an activity. The social marketing constant can be associated with a traditional advertising campaign or an event that triggers similar social media interest. Once people become exposed to an advertisement and decide to share the message, the social interaction constant may be viewed as the natural tendency of the social media network to advertise internally through posts, tweets, and likes without external influences and advertising. We shall call this behavior social media chatter. Integrating these changes for a social event-based situation, we find that

$$dX_t = \beta_1 u(t)[1 - X_t] + \beta_2[1 - X_t]X_t - \delta X_t. \tag{9.8}$$

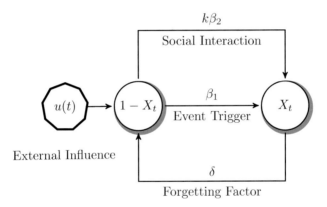

FIGURE 9.2: Flow diagram for an event-triggered social chatter model.

It is worth remarking that the control action $u(t)$ is connected to the effectiveness constant β_1 for the triggering event because only current events or active campaigning can be realistically used as control factors in this framework. A sufficient control factor acting directly on the social interaction term β_2 would be troublesome. This difficulty primarily stems from the diversity of the impulses and tendencies of social media users, making the response of any given society difficult to foresee. Trying to exert adequate control over said community social interactions outright would require a control factor custom-designed to each network group and is therefore practically unattainable for this general proposed modeling framework. However, many dynamical models implement both internal and external control influences ($u(t)$ is used for this model), which have an aggregated control action impact on the system. This is because external forces on social media (and many other systems) can be modeled on a macroscopic dynamical order by looking at each outlet in aggregate [73]. In other words, applying control to a reasonably similar group can be simulated using methods effective at controlling an average individual of that group. Liberal voters in the same economic class and city, for example, will respond fairly predictably to certain events or policies. Control over that group can be exercised with the right social campaigns, memes, or speeches. A similar macroscale approach is taken in other literature by using a consolidated trigger on all the dispersion forces [69]. Additionally, with the decay factor in place, topics will ultimately dissolve into background social media dialogue as people generally become disinterested with news over time.

However, the model considered in Equation 9.8 does not take into account the connectedness factor k used in the mean-field epidemiology-based models. Using a homogeneous mean field approach, the model can be adjusted as follows:

$$dX_t = \beta_1 u(t)[1 - X_t] + k\beta_2[1 - X_t]X_t - \delta X_t. \tag{9.9}$$

A summary of the variables and parameters for the event-based social media chatter model can be found in Table 9.4.

TABLE 9.4: Event-based social media chatter model variables and parameters

Term	Meaning
X_t	Spread of social media topic at time t
$u(t)$	Control action event at time t
β_1	Social marketing campaign constant
β_2	Social interaction constant
δ	Decay constant
k	Social network connectedness factor

9.4.1   Socio-Equilibrium Threshold

For a social media response to an story or incident to capture attention and become broadly shared, it must be capable of sustaining itself for a time upon the ending of external and forced triggering activities. When the control $u(t)$ becomes zero, the event or promotion has terminated. What remains are the second two terms of Equation 9.9. When the sum of these two terms is equal to zero, the system is in an equilibrium state. When

$$dX_t = k\beta_2[1 - X_t]X_t - \delta X_t = 0$$

that means,

$$X_{eqb} = 1 - \frac{\delta}{k\beta_2} \tag{9.10}$$

is the equilibrium level of social media chatter after the end of the triggering event. Commonly, this level will be low in the middle of a comparable triggering occasion and can be understood qualitatively as online remarks or discussions of a topic. Similarly, it can acquire attention for an event indirectly due to an outside cultural, political, physical, or social occurrence. Technology companies, for example, are often discussed on social media for their minor products, policy decisions, or scandals (at equilibrium). These same companies are much more hotly debated and discussed online when a major announcement is made concerning a highly anticipated consumer product such as a phone or video game console.

9.4.2   Simulation and Discussion

Now that we have an understanding of the event-triggered social media chatter model and the concept of a socio-equilibrium threshold, we can simulate the model under various conditions to achieve visual insight into how it operates. Beginning with Equation 9.9, we first assume very few people are aware of the topic to be socially marketed and there are no outside advertisements (no control actions) used to spread the information throughout social media. A simulation of this scenario is shown in Figure 9.3.

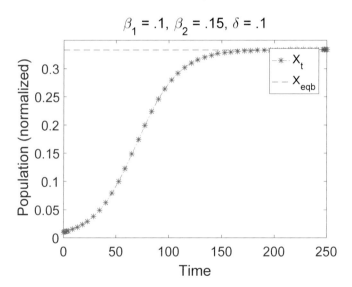

FIGURE 9.3: Event-triggered social media chatter model without control.

Notice that without active control, social media groups will naturally reach an equilibrium point (X_{eqb}) of interest in the product, social movement, meme, etc. However, reaching this point purely though social media "chatter" takes a significant amount of time and never reaches widespread attention. This is where social media marketing comes into play.

The goal of social media marketing is to increase peoples' attention and interest beyond the natural equilibrium point through the control action of spending resources on ads, online reviews, channel sponsorship, official Tweets, etc. If purposely influenced, this control is exerted intermittently as interest levels begin to dip, giving the social media chatter a "boost" and keeping general enthusiasm and discussions high. In the case chatter about general topics, significant events related to that topic that arise intermittently will give a similar control action boost, intentionally or not. Figure 9.4 gives a simulation of intermittent social marketing control using the social marketing chatter model.

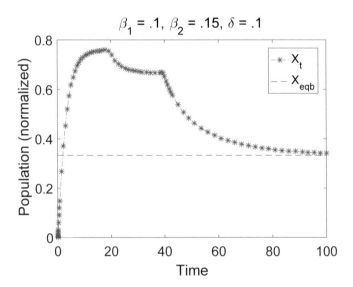

FIGURE 9.4: Event-triggered social media chatter with intermittent control.

With an intermittent control simulation and the same parameters, a strong level of control is used initially until time $t = 20$. As a result, over three-quarters of the population is aware of the information and spreading it. After this point, some control is used (fewer ads and marketing tactics than initially) and interest begins to die down, though it may still be widely discussed on social media. At time $t = 50$, all control activities are halted and the social media chatter slowly reaches the original (marketing-free) equilibrium level X_{eqb}. Notice that the time to reach an equilibrium of social media chatter is greatly reduced when any marketing is used to build up initial interest in a product or topic. For the sake of completeness, Figure 9.5 gives the control action profile used over time during intermittent control for this example. Control methods and applications will be discussed in detail in future chapters.

The next chapter will present case studies that examine several key social media information spreading models discussed thus far.

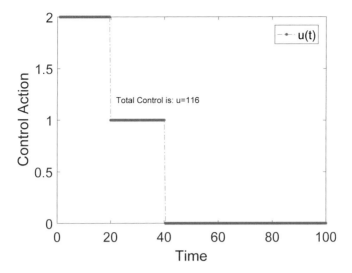

FIGURE 9.5: Intermittent control over time for a simulated event-triggered social media chatter.

```
                    Code 9.1: Event-triggered Social Chatter Model
1  %This script will calculate and plot the
       trajectory of spreaders in a
2  %social marketing model with variable control
3  %dXt = [B1*(1-x1).*u + B2*(1-x1)*x1' - d*x1];
4  clear;
5  to = 0;   % initial time
6  tf = 100; % final time
7  yo = [0]; % starting spreaders
8  x_eqb = .333; % equilibrium point (1-a\b1)
9  [t,u] = ode45('ypMarketing_SA',[to tf],yo);
10
11 figure(1);
12 plot(t,u(:,1),'-.r*'); hold on % plot spreaders
13 plot(t,ones(size(t))*x_eqb,'b--'); % plot
       equilibirum line
14 title('\beta_1 = .1, \beta_2 = .15, \delta = .1',
       'FontWeight','Normal')
15 xlabel('Time')
16 ylabel('Population (normalized)')
17 legend('X_t','X_{eqb}')
```

```
18  ax = gca;
19  ax.FontSize = 16;
20  hold off
21
22  time = 1:.5:100;
23  u = zeros(1,length(time));
24  u(time >=0 & time <=20) = 2;
25  u(time >=20 & time <=40) = 1;
26  u(time >=40 & time <=60) = .5;
27  u(time >=60 & time <=100) = 0;
28
29  %Calculate total control
30  u_total = trapz(u)
31
32  figure(2);
33  plot(time, u,'-.r.','MarkerSize',10);
34  title('\beta_1 = .1, \beta_2 = .15, \delta = .1',
        'FontWeight','Normal')
35  s = strcat('Total Control is: u=',num2str(u_total)
        );
36  text(22,1.2,s);
37  xlabel('Time');
38  ylabel('Control Action');
39  legend('u(t)');
40  ax = gca;
41  ax.FontSize = 16;
```

Code 9.2: Model Implementation Function

```
1  function ypMarketing = ypMarketing_SA(t,y)
2  %This function outputs the dynamics to be analyzed
3  %   Detailed explanation goes here
4  delta = .1; % forgetting factor
5  b1 = .1; % spreading constant
6  b2 = .15; % social spreading
7  ut = ctrl(t); % control profile
8  X_eqb = 1-(delta/b2); %equilibrium state of social
        interaction
9  ypMarketing(1) = b1*ut*[1-y(1)] + b2*[1-y(1)].*y
        (1)-delta*y(1);
10 ypMarketing = [ypMarketing(1)]';
11
12 %nested function to do time varying control action
13   function ut = ctrl(t)
14     if t< 20
15         ut = 2;
16     elseif 20<=t&& t<40
17         ut = 1;
18     else ut = 0;
19     end
20   end
21
22
23 end
```

9.5 Exercises

1. Discuss the need for an 'event triggered social media chatter model' and a 'social media chatter model'. Also, shed light on its connection with traditional marketing models discussed in the text.

2. Identify four case studies where an event caused social media chatter, i.e. a trending topic was created because of an event.

3. Marketing is only one of many well-established methods of actively swaying public opinion. Select another traditional method and describe how it both succeeds and requires changes for usage in a social media environment.

4. Use the simulation code provided in this chapter to simulate a social marketing event of your choice. It can be a social cause, a product, a political event, etc. Choose new parameters and calculate the equilibrium point X_{eqb}. Set various control strengths for $u(t)$ at different times and notice how the social marketing campaign evolves. Describe this evolution in terms of control and social media chatter. Plot the results.

10

Case Studies

> *There does not exist a category of science to which one can give the name applied science. There are sciences and the applications of science, bound together as the fruit of the tree which bears it.*

Louis Pasteur, *Revue Scientifique*, 1871

A case study is a descriptive and exploratory analysis of a real-world person, group, or event. Case studies are critical in testing ideas in areas, such as social sciences, medicine, and more because they can ultimately give a comparison between theory and reality. Additionally, case studies can be used to correct or tune existing beliefs within a subject matter.

Practically speaking, a case study involves collecting data on a person, group, or event and analyzing the resulting data. For our purposes, case studies are essential because they can help us validate social media information spread models. Given the great diversity and nuance of online interactions, privacy requirements, and the probabilistic nature of examining human interaction as a whole, case studies provide valuable insight for researchers. Once a case study has been run, the results can be compared both qualitatively and quantitatively to existing models using a variety of instance-based techniques.

As such, the focus of this chapter is on case studies, specifically within the context of social media. It begins with addressing some considerations and methods for case study selection, using Google Trends as a relatable and easily accessible example. Next, four different case studies are presented: 2017 mass shootings, the #MeToo social movement, the 2018 Golden Globe Awards, and viral internet debates. Each case study is applied to one of the information spread models presented in this text toward the end goal of verifying the model and its context-appropriate usage to the study of information spread over social media. For each of these case studies, data acquisition methods using the Twitter platform are first discussed. The methods used to estimate the model parameters are presented and calculated. Using the estimated parameters, model-based simulation results are compared to the actual data collected to help validate each model. A brief discussion of the results concludes each case study.

10.1 Selecting Case Studies

But where does one begin? Certainly, a case study subject must be chosen, and data must be collected. Data collection can be a lengthy and potentially costly endeavor without first pinpointing the subject of the study. Luckily, there are a number of tools to at least help a researcher obtain a preliminary sense of the data on a person, group, or event as a good first step. One such tool, for example, is the popular *Google Trends*. Google Trends compiles and organizes real Google searches, allowing you to enter a search term and see the common and trending terms related to those you entered. In principle, people are searching for things they find curious, news, and events they wish to know more about, topics being discussed as a culture, and more. Yes, searches alone do not give researchers a good or detailed picture of information spread. Still, it might provide a clue as to a starting point, before resorting to web scraping, purchasing data, or other methods of obtaining data. Additionally, trending searches can give hints of topics that should have significant amounts of interest and data to eventually conduct a case study of adequate population size (the more data points, the better). A cursory "trending" search shows that one popular search trend in the past several years has been the Apple iPhone. It is unsurprising, as the iPhone has been a trendy consumer device since its initial release in 2007. Figure 10.1 shows the Google Trends search for iPhones from 2004 to 2020.

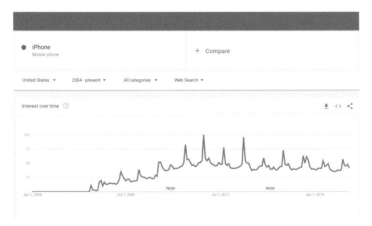

FIGURE 10.1: Google Trends search: "iPhone". Data source: Google Trends (https://www.google.com/trends).

The graph shows what we might reasonably expect: several peaks that coincide with the release of each new iPhone model, with a general popularity trend and stabilization over time. Each cycle also has a much smaller, secondary peak. What could this be? It seems to be present in some form during every period. Let's take a closer look via another Google Trends search. At the time of writing this text, the iPhone 11 was the most recently announced iPhone; we will search based on the year of its release in 2019, as shown in Figure 10.2.

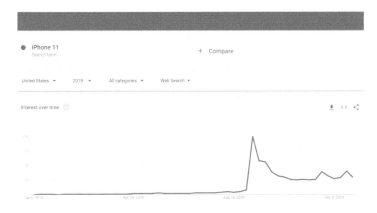

FIGURE 10.2: Google Trends search: "iPhone11". Data source: Google Trends (https://www.google.com/trends).

It is immediately apparent that there are two smaller peaks after the first announcement peak in 2019. Why not a single peak like in every other year? The most likely explanation (observable when examining each year more closely) is that both peaks are added together as a single point for a longer time scale. This is something that must be kept in mind when examining raw data. The proper "data resolution" can have a significant impact on our initial perceptions and analysis.

By researching the dates at which these two sub-peaks occur, we can infer that they are a result of U.S. holiday sales (Black Friday sales and Christmas sales, specifically). To support this hypothesis, an arbitrary iPhone model is chosen (iPhone 7 in this case), as shown in Figure 10.3.

Since the cyclical data and dates seem to generally align for both years (and can be easily reproduced for other years as well), we can check if the data and our offered explanation connect with any existing or proposed information spread models. We will now proceed with a more detailed and accurate data collection method, such as Twitter scraping or data purchasing.

In the next section, we present four case studies to demonstrate how a case study on social media information spread might be performed and how these case studies explain and validate the social media information spread models presented earlier. Note that other methods can be designed to fit a specific person, group, or event being researched, but these examples are hopefully good inspiration for designing other research methods. The four case studies are divided into two sets. The first set will use social marketing models in two different scenarios, while the second set focuses on epidemiology-based

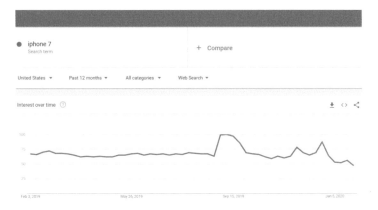

FIGURE 10.3: Google Trends search: "iPhone7". Data source: Google Trends (https://www.google.com/trends).

information spread models.

Data Sources: It is helpful to use data from participatory social media outlets such as Facebook when analyzing large-scale social behavior, sentiment, and communities [74, 75]. These outlets, however, are closed, friend-focused, and usually do not allow for consistent labeling or tagging functions to collect posts on a particular topic. For this reason, it can be challenging to observe the actual spread of information on platforms such as Facebook. Twitter, in contrast, requires concise messages due to its character limits, and offers an organized and well-utilized hashtag system for collecting messages by topic. Thus, for sentiment identification and analysis, Twitter is a superior data collecting source [76]. We selected Twitter as the sole social platform for the case studies in this text because it provides analysis tools and APIs to evaluate the relevancy of a selected hashtag. In the following case studies we examine the number of tweets using a particular hashtag at a given time to observe how the social campaigns develop and measure the intensity of the debate.

10.2   Case Study-1: 2017 Mass Shootings

Numerous events may trigger social media chatter, e.g. an advancing hurricane, a mass shooting, a favorite sporting event, etc. Figure 10.4 shows three event-triggered hashtags that trended in the year 2017: *#Shooting*, *#Hurricane*, and *#Eclipse2017*. The figure shows a stacked plot of the normalized amount of daily tweets for each of the three hashtags and how each evolves in usage over time. The peaks of each hashtag are marked with their accompanying major related events. On quick visual examination, all three hashtags appear to be good options for an event-based case study. Each hashtag has peaks in Twitter interest that directly correlate with the date of the corresponding major event triggers. From here, one hashtag must be evaluated based on its merits and used for our case study.

Sadly, there have been several instances of mass shootings in the United States during the year of interest, supported by extensive data on shooting-related events with which to conduct this research. By analyzing shooting-related Twitter data, insight is gained into the feelings and interests of users on social media directly following each of these mass shootings. People may hold very different opinions on gun violence and gun control and possibly be highly emotionally engaged in debating the topic. Compare this to the resulting emotional engagement (or lack thereof) from a natural phenomenon such as an eclipse. The chance for lasting discussion and interest is very strong for a case study based on mass shootings in contrast to hurricane and eclipse tweets. Here, we examine the case of mass shootings in the United States during the year 2017. The event-triggered social media chatter model discussed in the previous chapter is a good candidate as an appropriate model of the social media discussion following an unfortunate mass shooting incident.

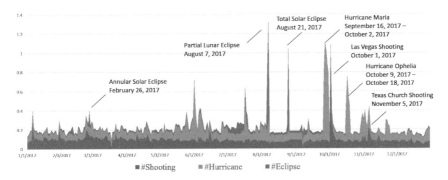

FIGURE 10.4: Normalized daily tweet volume of event triggered hashtags *#Shooting*, *#Hurricane*, *#Eclipse2017*. Courtesy of [69].

While there is no all-around acknowledged definition, the Investigative Assistance for Violent Crimes Act of 2012 characterizes mass shootings as shootings including at least three killings in a specific episode [77]. These events have been steeply increasing in recent years. A Wall Street Journal article saw that the amount of occurrences from 2000 to 2015 has dramatically increased from the first to the second 50% of the year run. To further aggravate the seriousness of the issue, 80% of the deadliest shootings in American history have happened in the previous five years. In 2017, the number of fatalities from mass shootings far exceeded those in any prior year in the United States [78]. Thus, we will concentrate on 2017 American mass shootings as a case study to show the adequacy of the proposed model for event-triggered social media chatter [69]. To research the case study further, different hashtags identified with mass shootings (for example, *#NRA* and *#GunControl*) are inspected separate from *#Shooting*. The everyday tweet amount of these three hashtags appears in Figure 10.5. The shootings in Las Vegas, Nevada, on October 1, 2017 (58 fatalities) and Sutherland Springs, Texas, on November 5, 2017 (26 fatalities) are indicated in the figure.

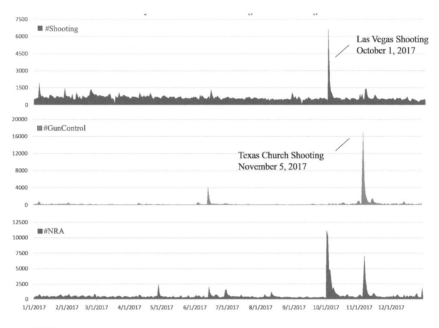

FIGURE 10.5: Daily tweet volume of shooting-related hashtags *#Shooting*, *#NRA*, *#GunControl*. Courtesy of [69].

10.2.1   Data Acquisition

We can validate the proposed model using Twitter data, which consists of the number of tweets using selected mass shooting-related hashtags. For this case study, we collected tweets made between January 1, 2017 and December 31, 2017 using a publicly available Twitter scraper. In total, we scraped 692,401 tweets for shooting-related hashtags. Table 10.1 breaks down the total number of scraped tweets for each hashtag.

We must note here that not every tweet containing the selected hashtags was collected by the scraper. The missing tweets may be the result of a few different limitations in the scraping tool. First of all, Twitter's application program interface (API) limited the retrieval of all tweets. Additionally, users could choose to adjust their privacy settings to protect the status of their

TABLE 10.1: Number of tweets scraped.

Hashtag	Scraped Tweets
#*Hurricane*	129,684
#*Eclipse*	29,981
#*Shooting*	151,343
#*NRA*	255,674
#*GunControl*	125,719
Total Tweet Count	692,401

tweets, disabling our ability to scrape them. Finally the architecture design of the scraping tool may also result in the retrieval of void data for some time periods.

The death data for the mass shootings was sourced from Mass Shooting Tracker [79], a crowd-sourced and publicly available database of U.S. mass shootings.

10.2.2   Parameter Estimation

The simulation of a nonlinear coupled ordinary differential equation (ODE) system is typically performed by suitable ODE solvers such as ODE45 and ODE15 in MATLAB. In contrast, the parameter estimation problem is cast as an optimization problem that is much trickier and can be categorized into global or local optimization methods. Global optimization methods include random search, adaptive stochastic processes, evolutionary computation, and clustering techniques. Although the global methods are relatively stable, they come at a high computational cost. Local optimization methods, in contrast, including Newton methods and quasi-Newton methods, are computationally efficient, but they do not guarantee convergence to global minima. In the case of parameter identification in ODEs, the problem of convergence to local minima is predominant if the so-called initial value approach is considered.

For the completeness of this text, we will present a simple method for parameter estimation from [80]. This is a wide and comprehensive topic, and for further reading into the topic, readers are directed to [81, 82], and [80].

Consider a general model for parameter estimation in the linear parameterization form

$$\mathbf{y}(t) = \mathbf{W}(t)\mathbf{a}, \tag{10.1}$$

where $\mathbf{y}(t)$ is the n-dimensional output vector, $n \times m$ dimensional $\mathbf{W}(t)$ is the signal matrix, and $\mathbf{a}$ is the m-dimensional vector of unknown parameters that need to be estimated [80]. Equation (10.1) is a linear equation in terms of the unknown $\mathbf{a}$ and $\mathbf{y}(t)$ and $\mathbf{W}(t)$ are known. In this case, the least-squares method is suitable to estimate the model parameters by minimizing the squared error between the observed and the estimated values.

However, the model, in this case, is a nonlinear differential equation (Equation 9.8). The dynamics of nonlinear systems can be expressed as a linear model taking the shape of Equation 10.1 by following a set of transformation steps as described in [80]. The method is detailed below by taking an example of a first-order differential equation.

Consider a first-order model

$$\dot{y} = -a_1 y + b_1 u, \qquad (10.2)$$

where model parameters a_1 and b_1 are to be estimated. The numerical computation of the derivative of y is not advisable due to error propagation. To avoid the derivative computation, both sides of the equation are multiplied by $1/(\rho + \lambda_f)$, where ρ is the Laplace operator and λ_f is a known positive constant. This step is also known as the filtering operation.

Rearranging we obtain

$$y(t) = y_f(\lambda_f - a_1) + u_f b_1, \qquad (10.3)$$

where $y_f = y/(\rho + \lambda_f)$ and $u_f = u/(\rho + \lambda_f)$. Now, the resulting Equation (10.3) is in the linear parameterization form and the two unknowns $(\lambda_f - a_1)$ and b_1 can be estimated using the minimum least squares method.

We are now ready to proceed with the case study. Recall the event triggered social media chatter model from the previous chapter:

$$\dot{X}_t = \beta_1 u(t)(1 - X_t) + \beta_2(1 - X_t)X_t - \delta X_t, \qquad (10.4)$$

where we take X_t as the daily volume of tweets containing a shooting-related hashtag, such as *#Shooting, #NRA, #GunControl*. And approximate $u(t)$ as the number of deaths due to mass shootings in the United States in 2017. We make an assumption and equate the event trigger ($u(t)$ in the model) to the number of deaths resulting from mass shootings. Although the assumption is crude, it will serve the purpose of demonstrating the model applicability. We apply the discussed procedure for parameter estimation and evaluate the model performance by computing the coefficient of determination (r-squared value), mean squared error (MSE), mean absolute error (MAE), and mean error (ME). For *#Shooting*, the parameters β_1, β_2, and δ are estimated to be 0.597, 1.196, and 1.175 with a r-squared score of 0.560. The equilibrium point X_{eqb} indicates a tweet volume level at which the hashtag settles down in the absence of external influences. In this case study, the equilibrium point X_{eqb} is estimated to be at 0.018. When compared to the parameter estimation result of *#Shooting*, the models of *#NRA* and *#GunControl* demonstrate smaller r-squared scores. The results are shown in Table 10.2.

One way to interpret the found results is that *#Shooting* is a reasonably unbiased and objective hashtag. In contrast, hashtags such as *#NRA* and *#GunControl* are more likely used to express a political philosophy or a politically motivated stand. On the Twitter platform, *#Shooting* also has a more significant number of appearances in news headlines or reports. Its tweet

TABLE 10.2: Parameter estimation of shooting-related hashtags.

Hashtag	β_1	β_2	δ	X_{eqb}	R-squared
#Shooting	0.597	1.196	1.175	0.018	0.560
#NRA	0.059	1.097	0.556	0.493	0.329
#GunControl	0.083	0.961	0.405	0.579	0.309

amount reflects the event timeline more precisely. Additionally, the use of an appropriate hashtag is a subjective matter, and these hashtags might overlap in a single tweet or might appear separately. The analysis of why users select a particular hashtag is beyond the scope of this book. However, one can argue that the above-discussed factors help explain the better performance of the event-triggered model on *#Shooting* data.

10.2.3   Results and Discussion

For this case study, we saw that a persistent social media debate on mass shootings evolves online through a number of steps. First, an actual event materializes. In this case, it is a shooting. Awareness of the event spreads throughout various social media and news platforms until conversations and general interest on the subject reach a saturation level. Eventually, interest and talk of the event and the related consequences will inevitably end. In the example presented here (of a mass shooting) the debate is reasonably persistent and will never fully end, but it will still reach a point where only a few people are involved in the active posting and discussing of mass shootings. However, interest can see a resurgence in both general and broad interest if another trigger (mass shooting in this situation) restarts the message spreading cycle.

In order to assess the effectiveness of the models developed to describe this type of social media information spread, simulations were performed based on data collected from Twitter posts following actual mass shooting events. The values X_t, $u(t)$, and the simulation parameters were used as described previously.

The orange line plot shown as the top half of Figure 10.6 shows the simulated information spread on social media. The blue line plot displays the actual Twitter chatter for the *#Shooting* hashtag. Finally, the bottom half of the figure shows the 2017 mass shooting event triggers by date based on estimated number of fatalities. To support the model's validity, correlations must be shown between the real-world Twitter data and the mathematical model simulation. As was expected qualitatively, the simulated trends show a steep increase, followed by a prolonged recession of information spread (in the form of tweets) upon the occurrence of each event. This pattern is consistent with the model's concept. Event-triggered social media activity develops as people have a strong initial impulse to tweet, post, or otherwise share new and noteworthy events. Still, the interest in circulating that information degrades over time until another mass shooting leads to rejuvenated public outrage, concern, opinions, and general discussion. Note that the peaks from the simulation of Twitter hashtag activity coincide strongly with the real number of deaths and of each mass shooting for any given date.

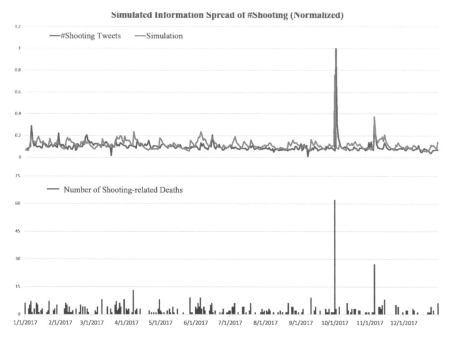

FIGURE 10.6: Simulation Results: Event-triggered chatter on *#Shooting* (normalized). Courtesy of [69].

In order to validate the model, quantitative measurement of the model's performance must be expressed. We computed the r-squared values, the MSE, the MAE, and the ME between the predicted number of the tweet and the observed number. The results are summarized in Table 10.6. Because both the number sets are normalized, the MSE values appear to be much smaller than the MAE values and the ME values. This can be explained by the fact that positive deviations and negative deviations can negate each other in aggregate. As shown in Figure 10.6, the variation between the tweet volume and the prediction simulation values are expressed by the absolute error. The MAE values of *#Shooting* (for this particular case) is 0.0364.

TABLE 10.3: Model validation result of hashtag *#Shooting*.

Hashtag	MSE	MAE	ME
#Shooting	0.0054	0.0364	0.0184

10.3 Case Study-2: The *#MeToo* Social Movement

In the next case study, we will examine the information diffusion model as applied to a social movement, cause, or craze scenario [69]. We utilize the same event-triggered social media chatter model used in the previous case study. Here, we define a social craze as a short-lived trend or fad. In the context of social media usage, a social craze can be expressed through memes, videos, or discussions including internet challenges, discussions about popular movies, social campaign participation, and more. The focus here will be on social movements. There are several recent hashtags that qualify as social causes including *#Resist*, *#BlackLivesMatter*, and *#MeToo* among others.

Figure 10.7 shows the tweet activity timeline for three recent social movement hashtags: *#Resist*, *#BlackLivesMatter*, and *#SexualHarassment*. In stacked plot form, it displays the normalized daily tweet volume by date for influential social campaign hashtags from October 2017 to March 2018 and events likely associated with spikes in Twitter activity related to the appropriate hashtags. This date range was chosen when a number of shocking sexual harassment accusations involving powerful celebrities (Louis C.K., Matt Lauer, and many others) came to light and dominated social media discussions and news reports. The social movement that resulted developed considerably during this period as new revelations, acknowledgments, and public attention took hold. In response, Twitter users began to share their own personal movement-related experiences and stories. By January 20, 2018, over one million people took part in street protests around the United States in the second annual Womens' March to express solidarity with victims and show support for the sexual harassment awareness movement.

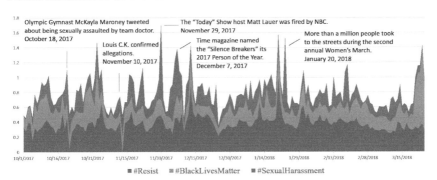

FIGURE 10.7: Normalized daily tweet volume of social campaign related hashtags. *#Resist*, *#BlackLivesMatter*, *#SexualHarassment*. Courtesy of [69].

This case study examines the rise of the social movement against sexual assault and harassment using the *#MeToo* hashtag. The movement was started by American activist Tarana Burke in 2006 and is widely believed to have been popularized by actress Alyssa Milano in 2017. The "Me Too" movement encouraged women who had been sexually harassed or assaulted to use the hashtag *#MeToo* when speaking on social media about their experiences. The movement led to a large number of tweets and Facebook posts over a short period of time [83]. On a high level, the tweets were information being spread on a large scale, acknowledging a serious social problem. This acknowledgment rapidly turned into a full-blown social campaign. Other hashtags associated with the *#MeToo* movement were also observed for this case study. These hashtags, *#TimesUp* and *#WhyWeWearBlack*, trended on the Twitter platform during the 75th Golden Globe Awards. During the televised awards show many actresses attending the ceremony wore black clothing in an attempt to spark widespread awareness of gender disparity issues and the problem of sexual harassment in the entertainment industry. Figure 10.8 shows the daily tweet volume activity of these three hashtags.

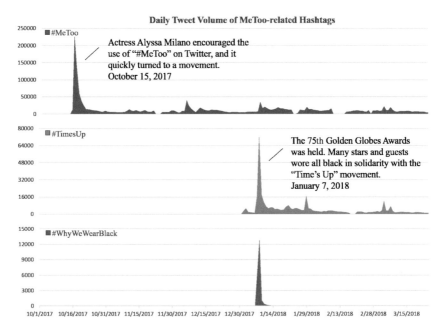

FIGURE 10.8: Daily tweet volume of MeToo-related hashtags *#MeToo*, *#TimesUp*, *#WhyWeWearBlack*. Courtesy of [69].

10.3.1 Data Acquisition

Tweets with relevant hashtags to this case study were scraped from October 1, 2017 to March 25, 2018. This date range was selected because the period best fits the popularity timeline of the *#MeToo* movement and adjacently related social campaigns. A total of 3,645,585 tweets were scraped. Table 10.4 shows the total number of scraped tweets for each hashtag. The shortcomings of the data acquisition remain the same as those explained in the first case study.

TABLE 10.4: Number of scraped tweets of social craze hashtags.

Hashtag	Scraped Tweet Count
#Resist	1,107,638
#BlackLivesMatter	179,353
#SexualHarassment	23,511
#MeToo	1,936,412
#TimesUp	376,801
#WhyWeWearBlack	21,870
Total Tweet Count	3,645,585

10.3.2 Parameter Estimation

Parameter estimation for the model described by Equation (9.8) is conducted similarly to the processes used and explained in the previous case study. For hashtag #MeToo, the parameters β_1, β_2, and δ are estimated to be 0.069, 1.418, and 1.416 with an r-squared score of 0.644. The equilibrium point X_{eqb} is estimated to be 0.002. The large volume of tweets for the hashtag #MeToo (1,936,412 tweets) contributed to a better mode performance as it relates to parameter estimation. In contrast, the modeling framework performed less effectively for the hashtag #WhyWeWearBlack, due to the availability of significantly less data (only 21,870 tweets). Table 10.5 gives the parameter estimation results of the three hashtags.

TABLE 10.5: Parameter estimation of MeToo-related hashtags.

Hashtag	β_1	β_2	δ	X_{eqb}	R-square
#MeToo	0.069	1.418	1.416	0.002	0.644
#TimesUp	0.040	1.239	0.480	0.613	0.459
#WhyWeWearBlack	0.036	0.272	-0.293	–	0.273

10.3.3 Results and Discussion

The #MeToo social campaign case study is an example of a social media conversation that began after an active social campaign to spread awareness for the cause. The cause was a popular and current topic that degraded over time. It is entirely possible that the debate will never wholly cease (though the hashtag association may change). However, it will still see diminishing discussion and reach a level in which only a comparatively small segment of social media is still involved in the active conversation of sexual harassment. Interest may still see a resurgence with a new or related social campaign against sexual harassment fueled by an influential tweet, comment, or action.

Actual data and simulation results are reasonably aligned for this case study. The top half of Figure 10.9 shows the model-based simulation of social media information spread (orange line plot) and the observed and collected Twitter data (blue line plot) of the *#MeToo* social movement from October 2017 to March 2018. External factors contributing to social awareness campaigns are shown on the bottom half of the plot. These external contributing factors were estimated based on the number of related major news outlet stories each day. The more breaking news stories of sexual harassment claims at one time the stronger the encouragement for it to be discussed on social media. The news database used to identify major breaking stories related to the hashtag was generated using The Chicago Tribune's "#MeToo: A Timeline of Events" report [84].

It is noteworthy that both the actual Twitter data and the timeline of the model-simulated tweets show a steep increase in interest which is followed by a slow decay after each associated major news story. This behavior is consistent overall with the modeling framework. Summarizing the behavior of a social campaign through social media, information spreads as people express an especially strong reaction to a developing movement coupled with a desire to share with others over social media. Still, interest in sharing and discussing that social movement information decays over time. Further events, social media "influencer" attention, political developments, or news stories lead to revived attention in the campaign and interest in spreading it. As with the previous case study, the simulation peaks identify these events and each (beyond the initial) aligns with renewed social media attention to the social movement.

We calculate the r-squared scores, the MSE, the MAE, and the ME between the model-predicted tweet volume and the observational data collected via tweets to measure model performance. The results of the model validation for *#MeToo* are displayed in Table 10.6.

TABLE 10.6: Model validation result for *#MeToo*.

Hashtag	MSE	MAE	ME
#MeToo	0.0123	0.0513	0.0285

The tweet volume of the model-predicted and the actual scraped tweet data are normalized. Because of this, the MSE values seem noticeably smaller than the MAE values and the ME values. However, positive deviations and negative deviations often balance each other (as can be viewed in Figure 10.9). The difference between the tweet volume and the predicted spread is expressed as the absolute error more favorably than other metrics. The MAE value of

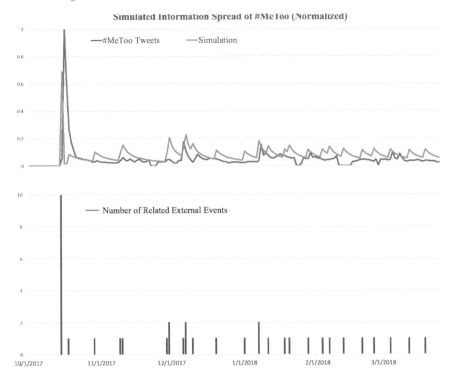

FIGURE 10.9: Simulation results: social chatter on *#MeToo* (normalized). Courtesy of [69].

#MeToo is 0.0513, and represents the average absolute error between the model prediction and the data collection observation results after all values were normalized.

10.4   Case Study-3: 2018 Golden Globe Awards

Usually, an event triggers a discussion that is then spread over social media. Political or social events, viral internet memes, and breaking news stories are all common examples of such discussion triggers. Single events such as these or other one-off occurrences can be described and analyzed using an Ignorant-Spreader-Recovered (ISR) type model for social media interactions. Due to the uniqueness and brevity of these triggers, they tend to initially attract considerable attention among social media users interested in spreading the information. Over time, there is a decay of popularity and interest in discussion and debate of the event as many of these people focus their attention on new topics and events they encounter in their interactions online.

10.4.1   Data Acquisition

Several hashtags were considered before ultimately deciding on the topic of the 2018 Golden Globe Awards, including the hashtags *#Hurricane* and *#Black-LivesMatter*. All of the considered topics are natural or cultural events that were trending in 2018. Each of these events led to energetic posts and discussions regarding catastrophic events and social inequality. Over the year, however, several of these topics had repeating event triggers (multiple hurricanes, protests, etc.) that would affect the online discussion and resurgence of interest in these topics. As such, we decided to select a single non-recurring popular event such as the 2018 Golden Globe Awards. Such an event can then be isolated as the primary source of discussions involving hashtags related to the event and will not be conflated with adjacent events involving the same broad topic. For more examples and information on the topic and hashtag selection process, consult [69].

The 2018 Golden Globe Awards granted an intuitive and useful case by which model information spread through use of the hashtag *#WhyWeWear-Black*. The hashtag refers to an instance during the award ceremony in which several actresses dressed in black gowns to show support for all women survivors of sexual assault. We utilized an open-source Twitter scraper to gather hashtag-related tweets for this case study [85]. This tool allowed us collect tweets under a specified hashtag over a desired range of dates. The total number of tweets collected with hashtag *#WhyWeWearBlack*, and the number of unique users are compiled in Table 10.7. The daily tweet volume for the hashtag during the week of January 5, 2018 to January 11, 2018 is shown in Figure 10.10.

TABLE 10.7: Dataset *#WhyWeWearBlack*.

Hashtag	Number of Tweets	Number of Users
#WhyWeWearBlack	21,870	13,613

We present a simple algorithm to rebuild the user tweets in the form of ISR groups when presented with a Twitter dataset. In the input dataset, each Twitter user must use a minimum of one tweet with the hashtag *#WhyWeWear-Black*. Upon first usage of the hashtag, the Twitter user class transitions from the ignorant to the spreader class group and is now actively spreading the message of interest. When the Twitter user tweets the same hashtag for the final time, the class of this user is assumed to change from a spreader to a recovered user as the user has stopped spreading the message for this specific subject. Since that user must be counted as only one person (and not once each tweet), similar tweets from the same user are not added to the spreader count.

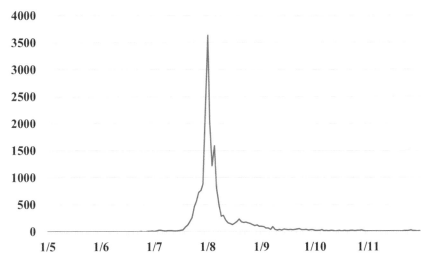

FIGURE 10.10: Tweet volume of #*WhyWeWearBlack*.

They must be represented as a single information-spreader, who is continuing to spread this message. The algorithm used to convert the Twitter data to ISR groups is shown in Algorithm 1. The estimated ignorant, spreader, and recovered group populations using this algorithm are shown in Figure 10.11.

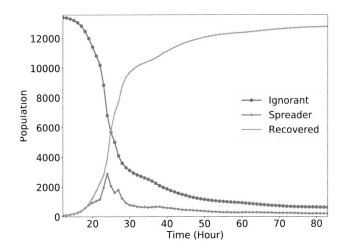

FIGURE 10.11: Reconstructed ISR groups for #*WhyWeWearBlack*. Courtesy of [63].

Through the construction process of Algorithm 1, each associated Twitter user underwent three stages in the information spread duration: ignorant, spreader, and recovered. It can be reasonably assumed that a Twitter user who terminated tweeting on a subject has lost interest in disseminating the information. This user is considered to have transitioned to the recovered group. When the exact instant a Twitter user began using a hashtag and discontinued using it is identified, the transition timeline of roles the Twitter users played (in aggregate) in spreading this hashtag is extracted, and each user group population count is determined.

Admittedly, some people were surely cognizant of this social campaign but chose not to post their opinions with the specified hashtags on Twitter's social media. These people are not captured in the ISR population groups by the discussed algorithm.

Algorithm 1 Constructing ISR groups from Twitter data

 1: $i \leftarrow$ *Ignorant Population*
 2: $s \leftarrow$ *Spreader Population*
 3: $r \leftarrow$ *Recovered Population*
 4: $t \leftarrow$ *First Hour of Data Collection*
 5: $t_{end} \leftarrow$ *Last Hour of Data Collection*
 6: **for** *each user* **do**
 7:     $t_0 \leftarrow$ *first time (hour) this user tweeted*
 8:     $t_1 \leftarrow$ *last time (hour) this user tweeted*
 9: **end for**
10: *loop*
11: **if** $t <= t_0$ **then:** $i = i + 1$, $t = t + 1$
12:     **goto** *loop.*
13: **else**
14:     **if** $t_0 < t <= t_1$ **then:** $s = s + 1$, $t = t + 1$
15:         **goto** *loop.*
16:     **else**
17:         **if** $t_1 < t <= t_end$ **then:** $r = r + 1$, $t = t + 1$
18:             **goto** *loop.*
19:         **else**
20:             **close** *loop;*
21:         **end if**
22:     **end if**
23: **end if**

10.4.2 Parameter Estimation

The parameters were estimated by following the steps in the previous case studies. Model parameters were estimated using the method of least squares by minimizing the squared error between the estimated (model-based) data and the observed (collected) data. See [80], [63], and [69] for supplementary details.

Recall the ISR Model for social media in Equation (7.11) and the ISR model for social media with decay in Equation (7.12). For the ISR Model for social media (without decay), there are two parameters to be estimated — β and γ. The parameter β is estimated to be 1.09. With 95% confidence, the value of this parameter falls in the range of 1.00 and 1.19. And the parameter γ is estimated to be 2.85, and the confidence intervals are 2.32 and 3.39. The confidence intervals of β and γ do not include zero value, which rejects the null hypothesis of no correlation. The r-squared score of β and γ are 0.90 and 0.64, respectively.

TABLE 10.8: Parameter estimation of the ISR model using #WhyWeWear-Black.

Parameter	Estimated Value	Standard Error	95% Confidence Intervals	R-Squared
β	1.09	0.05	[1.00 1.19]	0.90
γ	−0.10	0.62	[−1.34 1.15]	0.75

The second model, the ISR model with a decay factor, represents the act of forgetting information on social media and losing awareness and shall also be considered. The results of the parameter estimation for the collected data are shown in Table 10.9. It is worth remarking that the introduction of the decay factor has a major impact on the estimation of the spreader class and recovered class populations. In this case study, the estimated decay factor plays a prominent role in explaining the deterioration of the spreader group's population count, diminishing the role of γ.

TABLE 10.9: Parameter estimation of the ISR model with decay using #Why-WeWearBlack.

Parameter	Estimated Value	Standard Error	95% Confidence Intervals	R-Squared
β	1.09	0.05	[1.00 1.19]	0.90
γ	2.85	0.27	[2.32 3.39]	0.64
δ	0.56	0.11	[0.34 0.78]	0.75

R-squared values indicate that the model which considers a decay factor outperforms the model without it. The simulation results using these developed models are shown in Figure 10.12 and Figure 10.13.

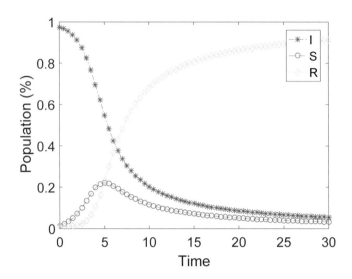

FIGURE 10.12: Simulated population of ISR groups for *#WhyWeWearBlack* (normalized, no decay considered).

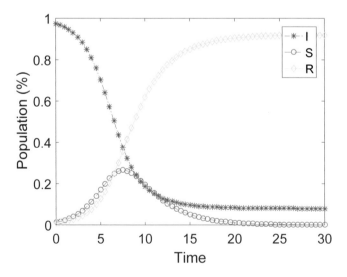

FIGURE 10.13: Simulated population of ISR groups for *#WhyWeWearBlack* (normalized, decay considered).

10.4.3 Results and Discussion

The results of the parameter estimation from the case study, displayed in Tables 10.8 and 10.9, show that the model effectiveness is enhanced with the inclusion of a decay factor. Because the modeling framework is intended to reasonably model the online behavior of users, it is expected that a model using a forgetting factor can capture the inherent features of online social communication better than one lacking a decay aspect. The simulation result comparisons in Figure 10.12 and Figure 10.13 further illustrate the disparity between the two models. Using the model with decay, the simulated spreader population declined more rapidly upon reaching its maximum value, which better follows the pattern of the reconstructed spreader population obtained using real data.

Again, several measurements between the simulated results and the reconstructed spreader population and the of model performance are calculated including the following: mean absolute percentage error (MAPE), MAE, MSE, and root mean squared error (RMSE). The model validation results for *#WhyWeWearBlack* for the ISR model are given in Table 10.10.

TABLE 10.10: Model validation result.

Hashtag	MAPE	MAE	MSE	RMSE
#WhyWeWearBlack	11.00%	0.08	0.01	0.09

10.5 Case Study-4: Viral Internet Debates

The last case study examines information spread for viral internet debates. Often, a video, image, meme, etc. attains mass appeal and ignites a social media discussion concerning the nature of the post or topic. Sometimes it can appear as a debate over sports such as who will win the next Super Bowl. On different occasions, these online discussions arise as a fervently debated advertisement and peoples' responses to it [86]. Debates like these are common in-person, on social media, and on the internet in general.

In this case study, we will consider the debate surrounding a photograph of an ambiguously colored dress posted to social media. A picture of the dress is shown in Figure 10.14. The environmental lighting of the dress resulted in an optical illusion. Some people saw the dress as blue and black, while to others it appeared white and gold [87]. For a brief timeframe, the true color of the dress was discussed and immediately developed to viral status. The debate even spilled into conventional television news outlets. Individuals wound up in one of two camps concerning the dress, each attempting to persuade the opposite

FIGURE 10.14: The original *The Dress* picture.

side that their perspective was right. Eventually, the genuine dress from which the images started was found, and the dress's true color was reported as blue and black. After the discovery, the majority of those in the inaccurate color group stopped trying to "convert" the contrary side to their point of view. Such a scenario is a good case study candidate to validate the previously discussed ISCR model.

10.5.1   Data Acquisition

In this case study, hashtag *#BlueandBlack* and its opposing counter-hashtag *#WhiteandGold* were chosen to symbolize the conflicting perspectives of the dress color. Tweets were gathered from February 1, 2015 to March 31, 2015 [85]. The total number of user tweets and the number of unique Twitter users for each of the two hashtags are presented in Table 10.11. Altogether, 138, 547 tweets from this date range were utilized for simulating and validating the ISCR model. Figure 10.15 shows the tweet volume of hashtags *#BlueandBlack* and *#WhiteandGold*. Spreaders were defined as users who tagged their tweets with hashtag *#BlueandBlack*. Meanwhile, counter-spreaders were defined as those Tweeting under the *#WhiteandGold* hashtag. An algorithm similar to the one used in the previous case study was implemented to reconstruct the ignorant, spreader, counter-spreader, and recovered groups from the Twitter data. The ISCR group population dynamics are displayed in Figure 10.16.

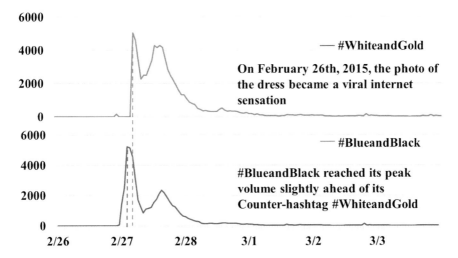

FIGURE 10.15: Tweet volume of *#BlueandBlack* and *#WhiteandGold*. Courtesy of [63].

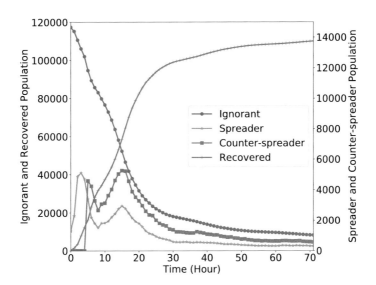

FIGURE 10.16: Reconstructed ISCR groups for *#BlueandBlack* and *#WhiteandGold*. Courtesy of [63].

TABLE 10.11: Dataset *#BlueandBlack* and *#WhiteandGold*.

Hashtag	Number of Tweets	Number of Users
#BlueandBlack	58,398	51,114
#WhiteandGold	80,149	67,868

Concurrence can also be expressed via retweets. On the Twitter platform, a retweet is the act of posting a tweet encountered on another person's timeline to the user's personal Twitter timeline. The quantity of retweeting activities each tweet pulls in from other Twitter users can also be an good measure of popular opinion. The dataset used in this case study, however, does not have retweet numbers connected with each unique tweet collected in this case study. This is a limitation of the data collection methods available and should be considered in future data collection attempts.

10.5.2   Parameter Estimation

Recall that we considered the ISCR model and its applications toward describing contentious information spread, mathematically expressed in Equation (7.13). The parameters for the ISCR model were estimated using the previously explained methods. For the two hashtags *#BlueandBlack* and *#WhiteandGold*, the parameters α, β, γ and μ are estimated to be 1.33, 0.90, 8.47, and 2.54. The relative difference in receptivity between the spreader and counter-spreader groups, $\omega_2 - \omega_1$ is estimated to be 47.18. Recall that the parameter $\omega_2 - \omega_1$ represents the transition rate of a counter-spreader to a spreader, measuring one's changed opinion in this instance. As seen in Figure 10.16, counter-spreaders consistently constitute a higher percentage of the total population. The parameter estimation results are shown in Table 10.12. And utilizing the result of the parameter estimation, Figure 10.17 shows the simulation of the ISCR population groups for this case study.

TABLE 10.12: Parameter estimation of the ISCR model using *#BlueandBlack* and *#WhiteandGold*.

Parameter	Estimated Value	Standard Error	95% Confidence Intervals	R-Squared
α	1.33	0.10	[1.13 1.52]	0.90
β	0.90	0.09	[0.72 1.07]	0.90
γ	8.47	1.33	[5.80 11.15]	0.59
μ	2.54	0.64	[1.24 3.83]	0.59
$\omega_2 - \omega_1$	47.18	9.33	[28.41 65.95]	0.35

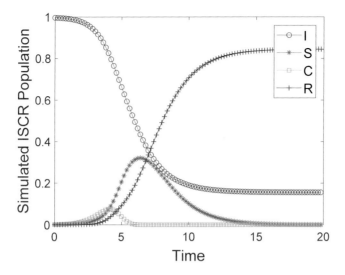

FIGURE 10.17: Simulated population of ISCR groups with *#BlueandBlack* and *#WhiteandGold* (normalized).

10.5.3 Results and Discussion

In this case study, a positive value of receptivity difference $(\omega_2 - \omega_1)$ indicates that throughout the social media debate between the spreader and counter-spreader groups, Twitter users supporting the hashtag *#BlueandBlack* were growing at a faster rate when compared to the growth rate of the *#White-andGold* counter-spreader group. Figure 10.16 and Figure 10.17 make the receptivity difference's influence on growth rates visually evident through the reconstructed ISCR population dynamics and the simulated population dynamics, respectively. Please note that the maximum population of spreaders is clearly higher than the maximum population of counter-spreaders.

Again, in order to measure the performance of the model four metrics are computed between the simulated results and the data-constructed spreader population: MAPE, MAE, MSE, and RMSE. The validation results for the ISCR model using hashtags *#BlueandBlack* and *#WhiteandGold* are summarized in Table 10.13.

TABLE 10.13: Model validation result.

Hashtag	MAPE	MAE	MSE	RMSE
#BlueandBlack	10.12%	0.11	0.01	0.12
#WhiteandGold	17.68%	0.12	0.02	0.14

When we compare the results in Tables 10.10 and 10.13, it can be observed that the estimation performance of *#BlueandBlack* and *#WhiteandGold* was not as good as *#WhyWeWearBlack*.

There might be multiple reasons for the lower model accuracy. It may be the model itself that is trying to capture a complex social phenomenon too simplistically. The subject of these topics might have also affected the estimation performance. This is perhaps because the hashtag pair *#BlueandBlack* and *#WhiteandGold* are specific to the sentiment and opinion they were intended to express in the debate for each sub-group. In contrast, *#WhyWeWearBlack* is a relatively general hashtag from the previous case study and can be used to reference different features of the sexual assault and awareness movement.

Concluding remarks: The discussed case studies in this chapter provide an end-to-end framework from data collection to model validation concerning modeling social contagion and information diffusion. These case studies were sourced from published articles, and readers are encouraged to review [63] and [69] for in-depth analysis. These case studies serve as useful illustrations of real-world applications of the macroscopic modeling framework that was developed in the preceding chapters. However, some remaining limitations must be mentioned here. The restrictions on data acquisition include the consideration of only the Twitter platform to measure information spread and exclude other popular platforms such as Facebook or Instagram. Moreover, the data was not officially sourced from Twitter but scraped, gathering publicly available tweets. This may lack the retweet (or general re-posting) timeline of each gathered tweet, which can contain valuable information on message broadcasts over various social media platforms.

10.6   Exercises

1. What are the main considerations involved in choosing a case study to test a particular information spread model?

2. Try to obtain Twitter data for the two case studies described in the exercise section of Chapter 7. Use this data to validate your chosen model by following the steps described in this chapter.

3. After reviewing the case studies presented in this chapter, what information is important to include in the results and discussion sections? How should it be discussed in order to be both academically ethical and convincing to readers?

4. Describe how case studies help serve to validate a proposed model. List some of the model-to-data comparisons that must be made and key statistical metrics. Why are they important?

5. One method of model validation is to split collected data into two parts. The first part is used to determine model parameters. Those parameters are used to predict the remaining data points and the theoretical and actual values are compared. What are some of the advantages and weaknesses of this method?

Part III

Control Methods For Information Spread

11

Control Basics

> " *Be the change that you wish to see in the world.* "

Mahatma Gandhi

This chapter aims to develop a basic understanding of control systems from an intuitive point of view. In this chapter, we provide an informal introduction to control systems design by discussing a few day-to-day life examples involving control design, open and closed feedback control systems, single-input single-output, and multiple-input multiple-output control systems. Some fundamental steps for control system design are given to conclude the chapter. It is aimed at a reader who considers themselves a layman in control theory and wants to gain intuitive insights before diving deep and involving mathematical details. Readers with prior knowledge of control theory may choose to skip this chapter.

11.1 Introduction

Control systems engineering is tasked with understanding and controlling an entity or group of entities termed as a *system*. The objective is to control the system such that it performs the desired tasks and behaves as per the requirements. Often, modeling and control of the system go hand in hand. Effective modeling of the system helps in understanding the system behavior in a given set of conditions and inputs. This understanding, in turn, helps in designing suitable control strategies for the system. In the absence of an effective modeling framework of a system, control design reduces to just guesswork as we are not sure what the system response will be given the control input. In other words, the input-output relationship must be known to a reasonable extent. Figure 11.1 conceptually illustrates input and output to a system.

Control systems engineering is an interdisciplinary field that has existed for a long time. This field has broad applications in chemical, electrical, aeronautical, mechanical, and environmental engineering. Our systems, and their control design, are getting increasingly complex and interconnected with the advances of communication, sensing, and technology. Systems such as

FIGURE 11.1: Control input and output.

intelligent transportation systems, socio-technical systems, and information propagation on social media provide excellent examples of large-scale and complex systems that are often challenging to model and control.

The foundation of control theory lies in the linear systems theory, which views systems components using a cause-effect theory. Now, based on certain parameters, there may be many categorizations of control systems; among them is open-loop vs. closed-loop control systems, SISO and MIMO control systems, continuous-time and discrete-time control systems.

11.2   Open-loop and Closed-loop Control Systems

In an open loop control design, an input is estimated, which will cause the desired output when implemented using an actuator. A point to note here is that there is no feedback or sensing mechanism during the process or the actuation. Thus the control action is static in nature, and we can assess the actual vs. desired output only after the actuation is complete. Figure 11.2 illustrates a block diagram for an open-loop control system.

FIGURE 11.2: Open-loop control system.

For example, consider Figure 11.3 which illustrates **open-loop control** application in day-to-day life. Let us say that you have to toast your bread in a toaster. Now depending on how crisp you like your bread (desired output), you will have to adjust the timer based on your prior experiences (system knowledge and model). This knob adjustment is the control input you have provided to the toaster (actuator), which will apply the input and deliver the output in the form of the toasted bread. Now, since there was no sensing or modification of control input during the process (toasting), this falls under the open-loop control category. You will know the output only after the toasting process is complete, and the bread is out of the toaster. If you got an over toasted or burnt bread, you would probably need to rethink your control strategy. Better luck next time!

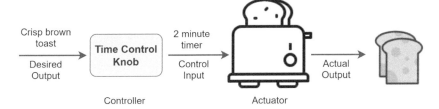

FIGURE 11.3: Bread toaster: open-loop feedback control system.

Unlike the open-loop control system, a closed-loop system performs output sensing and measurements as feedback during the process and alters the control action accordingly. Actual output is measured and compared with the desired output using a negative feedback loop, as shown in Figure 11.4. The feedback signal is subtracted from the desired output, and the difference (error) is fed to the controllers, which design the control action accordingly to continually reduce the error.

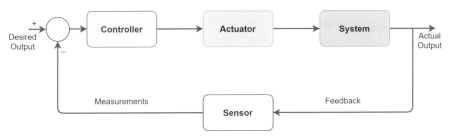

FIGURE 11.4: Closed-loop control system.

Consider a cruise control mechanism of a vehicle, as shown in Figure 11.5. This is an example of a closed-loop feedback control system as the system is continually measuring the actual speed and comparing it with the desired speed set by the driver. The difference of speeds is then fed to the vehicle's electronic control unit, which designs a control action for the actuators on the internal combustion engine to implement.

11.3 SISO and MIMO Control Systems

Control systems can also be categorized based on the number of inputs and outputs involved. Single input and single output or SISO are control systems having only one input and one output. In contrast, multiple inputs and multiple

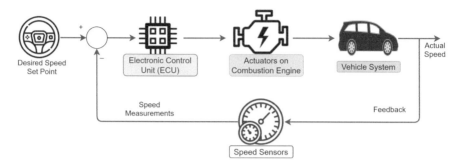

FIGURE 11.5: Cruise control: closed loop control system.

outputs or MIMO are control systems having more than one input and more than one output. For example, consider the spring-mass system described in Chapter 6. The system has only one external input $F(t)$, and only one output - position of the block $x(t)$. Hence this is an example of a SISO control system. Whereas, consider the three-mass-swinger shown in Figure 11.6. There are three forces as inputs $\mathbf{u} = [u_1 \; u_2 \; u_3]^T$ and three positions as outputs $\mathbf{x} = [x_1 \; x_2 \; x_3]^T$, making it a MIMO system.

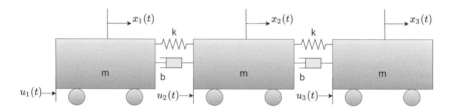

FIGURE 11.6: Multi-Input Multi-Output: three mass swinger.

11.4 Continuous-time and Discrete-time Control Systems

Control systems can also be categorized based on the type of signals involved. Continuous-time control systems have all the signals in a time continuum where any two points in time contain an infinite number of in-between points. In contrast, discrete-time control systems have all the signals in separate and distinct points in time, moving from one time period to the next. For example, a year of time can described as a continuous-time flow from one day, minute, second, millisecond, etc. to the next on an infinitesimally small scale. That same year can be viewed in discrete-time as a set of twelve explicit months. Historically, control systems were built using analog components, meaning that the control and input signals both were in continuous time. Whereas the advent of discrete or logical control components in parallel with computer-based control systems gave rise to the use of discrete-time signals.

11.5   Control System Design

The control system design is a class of engineering design problem. Just like any other design problem, it requires both system analysis and creativity. The engineering control design process involves multiple steps, as shown in Figure 11.7. The first step is the identification of control goals, which includes clarifying the specifications, defining the system, and identifying system variables, control variables, and key parameters. The second step involves defining system boundaries and establishing an appropriate system model. The third step involves control design and parameter estimation. And the fourth step involves performance evaluation using the designed control.

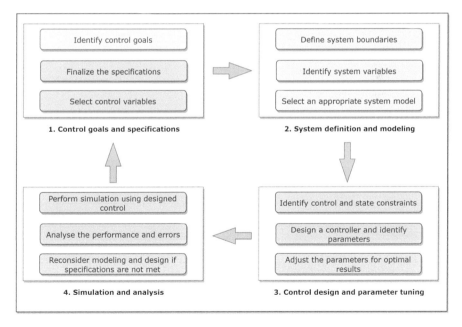

FIGURE 11.7: Steps for control system design.

12

Control Methods

> *When it is obvious that the goals cannot be reached, don't adjust*
> *the goals, adjust the action steps.*

Confucius

The focus of this book has been on understanding and modeling the spread of information on social media channels. However, the ultimate goal for any dynamical system is the ability to manipulate and drive it in the desired fashion. Similarly, for the considered system in this book, i.e. information spread systems, the goal is to utilize the developed modeling schemes for effective manipulation of the dynamical system. The study of designing manipulation schemes for dynamical systems is known as *control theory* — which involves concepts from applied mathematics and engineering. The success of any control method design is very much dependent on how well the developed model can capture the dynamics of the system. In that sense, control theory puts to test the approximations and assumptions during the modeling process.

Examples and applications of control theory are everywhere around us in the form of robotics, navigation of rockets, aircraft autopilot systems, adaptive cruise control of automobiles, heating and cooling systems in buildings, and many more. The formal control theory is believed to have been started by James Maxwell in 1868 during the analysis of the centrifugal governor. A more layman's introduction and intuitive explanation of control theory, including the two main types — open-loop and closed-loop control — was discussed in the previous chapter. The field has been evolving since its inception, and the methods range from a relatively simple PID controller to the more sophisticated optimal control methods. Depending on the dynamical systems and control objective, a controller must be chosen and designed carefully to meet the requirements.

In this chapter, we mathematically explain a few essential control methods in simple terms. However, the treatment is only cursory as needed for the following control application chapters. Readers are encouraged to follow the suggested references for an in-depth treatment of these control topics. The chapter begins with the design of a state variable feedback controller, including controller design, observer design, and an integrated design. A basic summary of PID controller design is also given. Optimal control methods are presented in more detail, covering performance measures, dynamic programming, and Pontryagin's minimization principle. A discussion concerning the application of optimal control methods in social media systems concludes the chapter.

12.1 State Variable Feedback Controller

State variables and state-space forms can be utilized to design a required control method for linear systems in the time domain. The control action $\mathbf{u}(t)$ is a function of all or some state variables. When all the states are utilized to build a controller, it is referred to as *full-state feedback controller*. Generally, as the measurements of all the state variables are not available, we need to build an observer to estimate the states that are not directly measured.

12.1.1 Full-state Feedback Control Design

The full-state feedback controller employs all the state variables to place the poles of the system at the desired location. Let the following state variable model represent the linear system

$$\left\{ \begin{array}{l} \dot{\mathbf{x}}(t) = \mathbf{A}\mathbf{x}(t) + \mathbf{B}u(t) \\ \mathbf{y}(t) = \mathbf{C}\mathbf{x}(t), \end{array} \right. \tag{12.1}$$

where $\mathbf{A}$, $\mathbf{B}$, and $\mathbf{C}$ matrices have the same meaning as described previously in Chapter 6.

Regulation Problem: The control objective is identified as a regulation problem when we require the states to become zero beginning from any given initial state. The design also ensures the system's internal stability while taking into account the desired transients. Assuming that the measurements of all state variables $\mathbf{x}(t)$ are available, the system control input for a regulation problem becomes

$$u(t) = -\mathbf{K}\mathbf{x}(t), \tag{12.2}$$

where $\mathbf{K}$ is the control gain to be designed. Using the control design in (12.2), the system in (12.1) becomes a closed-loop given as

$$\begin{aligned} \dot{\mathbf{x}}(t) &= \mathbf{A}\mathbf{x}(t) + \mathbf{B}u(t) \\ \dot{\mathbf{x}}(t) &= \mathbf{A}\mathbf{x}(t) - \mathbf{B}\mathbf{K}\mathbf{x}(t) \\ \dot{\mathbf{x}}(t) &= (\mathbf{A} - \mathbf{B}\mathbf{K})\mathbf{x}(t). \end{aligned} \tag{12.3}$$

The closed-loop system is stable if all its poles are in the left-half plane or, equivalently, if the roots of the associated characteristic equation are on the left-half plane. The characteristic equation of (12.3) is

$$\det(\lambda \mathbf{I} - (\mathbf{A} - \mathbf{BK})) = 0. \tag{12.4}$$

The solution of (12.3) is given as

$$\mathbf{x}(t) = e^{(\mathbf{A}-\mathbf{BK})t}\mathbf{x}(t_0), \tag{12.5}$$

here $\mathbf{x}(t) \to 0$ as $t \to \infty$ provided all the poles are located on the left hand side. If the system $(\mathbf{A}, \mathbf{B})$ is controllable, then we can surely obtain a $\mathbf{K}$, which places the poles of the closed-loop system on the left-half plane. For additional details on controllability and observability, readers are referred to consult [88] and [89].

Tracking Problem: The control objective is identified as a tracking problem when we require the states to track a reference signal $r(t)$. The steps to solve a tracking problem are similar to the regulation problem. Assuming that the measurements of all state variables $\mathbf{x}(t)$ are available, the system control input for a tracking problem becomes

$$u(t) = -\mathbf{K}\mathbf{x}(t) + Nr(t). \tag{12.6}$$

12.1.2 Observer Design

Usually, measuring all the states is neither feasible nor practical. If the system is fully observable, one can build an observer that can estimate the states using only the limited available state measurements. One such observer is the Luenberger observer described next. For the linear system described by

$$\begin{cases} \dot{\mathbf{x}}(t) = \mathbf{A}\mathbf{x}(t) + \mathbf{B}u(t) \\ y(t) = \mathbf{C}\mathbf{x}(t), \end{cases} \tag{12.7}$$

the Luenberger observer is given as

$$\dot{\hat{\mathbf{x}}}(t) = \mathbf{A}\hat{\mathbf{x}}(t) + \mathbf{B}u(t) + \mathbf{L}(y(t) - \mathbf{C}\hat{\mathbf{x}}(t)), \tag{12.8}$$

where $\hat{\mathbf{x}}(t)$ is the estimated state vector, and $\mathbf{L}$ is the Luenberger observer gain that one needs to be designed. The objective of the observer is to make estimation error go to zero, i.e.

$$\mathbf{e}(t) = \mathbf{x}(t) - \hat{\mathbf{x}}(t) \to 0 \quad \text{as } t \to \infty.$$

The time derivative of estimation error leads to the following error dynamics

$$\dot{\mathbf{e}}(t) = \dot{\mathbf{x}}(t) - \dot{\hat{\mathbf{x}}}(t)$$

which after further simplification becomes

$$\left\{ \quad \dot{\mathbf{e}}(t) = (\mathbf{A} - \mathbf{LC})\mathbf{e}(t), \right.$$

having the characteristic equation given as

$$\det(\lambda \mathbf{I} - (\mathbf{A} - \mathbf{LC})) = 0. \tag{12.9}$$

Provided that the system $(\mathbf{A}, \mathbf{C})$ is observable, one can design an appropriate $\mathbf{L}$ which places the poles of the closed-loop system on the left-half plane ensuring $\mathbf{e}(t) \to 0$ as $t \to \infty$. For additional details on controllability and observability readers are directed to [88].

12.1.3 Full-state Feedback Controller and Observer

The state variable compensator is obtained by coupling the designed feedback controller and observer. Since we do not have all the state measurements, we utilize the estimated state $\hat{\mathbf{x}}(t)$ in the control law as follows

$$u(t) = -\mathbf{K}\hat{\mathbf{x}}(t). \tag{12.10}$$

But the question arises if the independently designed controller and observer would still work when combined together. In other words, would the roots of the characteristic Equations (12.4) and (12.9) still lie in the left-half plane? The answer lies in the *separation principle* which states that full state feedback controller and observer can be designed separately and combined.

12.2 PID Controller

PID controllers have been extensively used in manufacturing processes and factories for very long time. PID stands for proportional-integral-derivative controller, and as the name suggests, PID is a combination of three controllers (see figure 12.1):

- Proportional Controller

- Integral Controller

- Derivative Controller

The three components have distinct roles addressing three different errors, making the overall controller versatile. The PID control is given as

$$u(t) = K_p e(t) + K_i \int_0^t e(\tau) d\tau + K_d \frac{de}{dt}, \tag{12.11}$$

where $u(t)$ is the control input and $e(t)$ is the system error, i.e. deviation from the reference trajectory $r(t)$.

The proportional controller rectifies the current error by being "proportional to the current error" in the system. Proportional gain (K_p) decides how quickly the controller reacts and how much steady-state error exists. A high (K_p) will lead the controller to react sharply to an error, which may decrease the steady-state error but, at the same time, create a risk for system instability.

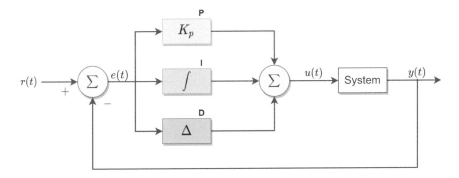

FIGURE 12.1: A block diagram of a PID controller.

The integral controller rectifies the past error, which may be accumulated by being "proportional to both the error and its duration". Thus the integral control gain (K_i) will reduce the steady-state error, but it may come at the cost of transient response.

The derivative controller rectifies future error and provides stability to the system by slowing down the controller's rate of change. The derivative control gain (K_d) stabilizes the system, betters the transient response, and reduces the overshoot.

There are several methods to tune the PID parameters — K_p, K_i, and K_d. These may be manual methods, the Ziegler-Nichols method, or the Cohen-Coon method. Readers interested in reading more about these methods and an in-depth explanation of PID control are referred to [90] and [91].

12.3 Optimal Control

The field of optimal control was first conceptualized in 1697, over 300 years ago. It started as a mathematical challenge by Johann Bernoulli — the Brachystochrone problem (see figure 12.2) [92]. Bernoulli's problem statement was as follows:

"Given two points A and B in a vertical plane, what is the curve traced out by a point acted on only by gravity, which starts at A and reaches B in the shortest time".

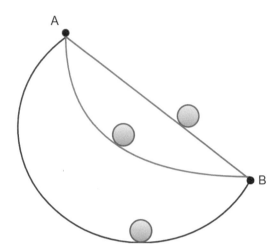

FIGURE 12.2: Brachystochrone problem.

Five mathematicians — Isaac Newton, Jakob Bernoulli, Gottfried Leibniz, Ehrenfried Walther von Tschirnhaus, and Guillaume de l'Hopital — each came up with their own solution which started the field of calculus of variations eventually leading to optimal control theory.

The idea behind optimal control is to find a system's ideal performance by defining the system criteria and optimizing performance measures. A control problem consists of a cost function or objective function, system dynamics, and constraints on state and control variables. The objective is to either minimize or maximize the cost function (made up of state and control variables) subject to certain constraints. Optimal control can be used in different disciplines, including biomedical devices, communication networks, or economic systems. Geometric or numerical approaches can be used to approach optimization problems. And, the nature of variables associated with optimization can be real numbers, integers, or combinations of both.

12.3.1 Performance Measure

A *performance measure* is a quantitative criterion that helps in evaluating the system performance under the designed control actions. The design of a performance measure is an important step in optimal control. Once the performance measure is identified, the goal of the optimal control design is to minimize or maximize that measure. Depending on the context, it may also be referred to as cost function, loss function, objective function, fitness function, utility function, or reward function — all having the same underlying meaning. The general form of performance measure is given as

$$\mathbf{J} = h(\mathbf{x}(t_f), t_f) + \int_{t_0}^{t_f} g(\mathbf{x}(\tau), \mathbf{u}(\tau), \tau) d\tau, \qquad (12.12)$$

where t_0 is the start time and t_f is the end time, h is the terminal cost at the end time t_f, and the function g captures the running cost.

As we will see in later chapters, optimal control is a natural technique for application in information spread, campaign management, and advertising. There are two main approaches to solve an optimal control problem — dynamic programming principle (DPP) and Pontryagin's minimization principle. We discuss both techniques in the following sections.

12.3.2   Dynamic Programming and Principle of Optimality

This segment presents a brief mathematical background for one of the methods used to solve an optimal control problem. Using the concepts of dynamic programming and the principle of optimality, we obtain the Hamilton-Jacobi-Bellman (HJB) equation, a partial differential equation (PDE), which establishes the basis for optimal control theory. For a formulated optimal control problem in terms of a system's state dynamics and an associated cost function, the concept of dynamic programming provides the solution in terms of a "value function", which minimizes or maximizes the cost function [93].

Assume the system dynamics are represented as

$$\dot{\mathbf{x}}(t) = \mathbf{a}(\mathbf{x}(t), \mathbf{u}(t), t), \qquad (12.13)$$

where $\mathbf{x}(t) = [x_1(t), x_2(t), \cdots, x_n(t)]^T$ is the state vector containing state variables and $\mathbf{u}(t) = [u_1(t), u_2(t), \cdots, u_m(t)]^T$ is the control vector containing control inputs.

The control objective is to find an optimal $\mathbf{u}^*$ that drives the system dynamics (12.13) such that the cost function

$$\mathbf{J} = h(\mathbf{x}(t_f), t_f) + \int_{t_0}^{t_f} g(\mathbf{x}(\tau), \mathbf{u}(\tau), \tau) d\tau \qquad (12.14)$$

is minimized.

The value function for the associated cost function is given as

$$\mathbf{J}^*(\mathbf{x}(t), t) = \max_{\substack{\mathbf{u}(\tau) \\ t \leq \tau \leq t_f}} \left\{ h(\mathbf{x}(t_f), t_f) + \int_{t}^{t_f} g(\mathbf{x}(\tau), \mathbf{u}(\tau), \tau) d\tau \right\} \qquad (12.15)$$

and it can be proved that a solution to equation (12.15) is obtained by solving the following HJB equation:

$$0 = \mathbf{J}_t^*(\mathbf{x}(t), t) + \min_{\mathbf{u}(t)} \{ g(\mathbf{x}(t), \mathbf{u}(t), t) + \mathbf{J}_\mathbf{x}^{*T} a(\mathbf{x}(t), \mathbf{u}(t), t) \}. \qquad (12.16)$$

Next, we define the Hamiltonian $\mathcal{H}$ as

$$\mathcal{H}(\mathbf{x}(t), \mathbf{u}^*(\mathbf{x}(t), \mathbf{J}_\mathbf{x}^*, t), \mathbf{J}_\mathbf{x}^*, t) = \min_{\mathbf{u}(t)}\{\mathcal{H}(\mathbf{x}(t), \mathbf{u}(t), \mathbf{J}_\mathbf{x}^*, t)\} \qquad (12.17)$$

then the HJB PDE can be rewritten as

$$\mathbf{J}_t^*(\mathbf{x}(t), t) + \mathcal{H}(\mathbf{x}(t), \mathbf{u}^*(\mathbf{x}(t), \mathbf{J}_\mathbf{x}^*, t), \mathbf{J}_\mathbf{x}^*, t) = 0, \qquad (12.18)$$

where the optimal control $\mathbf{u}^*(t)$ can be found by solving

$$\frac{\partial \mathcal{H}}{\partial \mathbf{u}} = 0.$$

The obtained optimal value of control $\mathbf{u}^*(t)$ is now substituted into equation (12.18), which creates a PDE in $\mathbf{J}^*$. This resulting PDE in $\mathbf{J}^*$ can be solved numerically or analytically (in special cases).

12.3.3 Pontryagin's Minimization Principle

Pontryagin's minimization principle is another prominent approach to optimal control based on the calculus of variations. The principle lays down a set of necessary (nut not sufficient) conditions required for optimality. In contrast, recall that the HJB equation provided sufficient conditions. Consequently, at best, the HJB equation can only guarantee local optimality within the plausible trajectories. The minimization principle alone may not lead to the conclusion that an obtained solution trajectory is optimal. However, it is considered valuable for obtaining potential optimal trajectories in several cases. Any candidate trajectory is not optimal if it fails to satisfy the necessary conditions laid by the minimum principle.

The minimization principle is generally formulated in terms of adjoint variables and a Hamiltonian function. If the system dynamics is given by

$$\dot{\mathbf{x}}(t) = \mathbf{a}(\mathbf{x}(t), \mathbf{u}(t), t) \qquad (12.19)$$

then the optimal control $\mathbf{u}^*$ which drives the system to minimize the cost function

$$\mathbf{J} = h(\mathbf{x}(t_f), t_f) + \int_{t_0}^{t_f} g(\mathbf{x}(\tau), \mathbf{u}(\tau), \tau) d\tau \qquad (12.20)$$

follows certain necessary conditions. First we define a Hamiltonian $\mathcal{H}$ as

$$\mathcal{H}(\mathbf{x}(t), \mathbf{u}(t), \mathbf{p}(t), t) = g(\mathbf{x}(t), \mathbf{u}(t), t) + \mathbf{p}^T(t)[\mathbf{a}(\mathbf{x}(t), \mathbf{u}(t), t)], \qquad (12.21)$$

where $\mathbf{p}(t)$ denotes an n-dimensional vector of adjoint variables. Now the necessary conditions so that $\mathbf{u}^*$ becomes the optimal control are:

$$\dot{\mathbf{x}}^*(t) = \frac{\partial \mathcal{H}}{\partial \mathbf{p}}(\mathbf{x}^*(t), \mathbf{u}^*(t), \mathbf{p}^*(t), t)$$

$$\dot{\mathbf{p}}^*(t) = -\frac{\partial \mathcal{H}}{\partial \mathbf{x}}(\mathbf{x}^*(t), \mathbf{u}^*(t), \mathbf{p}^*(t), t) \left.\right\} \quad \text{for all } t \in [t_0, t_f]$$

$$\mathcal{H}(\mathbf{x}^*(t), \mathbf{u}^*(t), \mathbf{p}^*(t), t) \le \mathcal{H}(\mathbf{x}^*(t), u(t), \mathbf{p}^*(t), t)$$

$$\text{for all admissible } \mathbf{u}(t)$$

and the boundary conditions are provided by

$$\left[\frac{\partial h}{\partial \mathbf{x}}(\mathbf{x}^*(t_f), t_f) - \mathbf{p}^*(t_f) \right]^T \partial \mathbf{x}_f$$

$$+ \left[\mathcal{H}(\mathbf{x}^*(t_f), \mathbf{u}^*(t_f), \mathbf{p}^*(t_f), t_f) + \frac{\partial h}{\partial t}(\mathbf{x}^*(t_f), t_f) \right] \partial t_f = 0. \quad (12.22)$$

Note that the formulation is in terms of a coupled nonlinear ODE system, unlike the HJB approach, in which we are required to solve PDEs. For more rigorous treatment and detailed explanation of dynamic programming, calculus of variations, and optimal control theory, we recommend *Optimal Control Theory: An Introduction* by Donald E. Kirk [93]. The applications of optimal control theory are widespread, and this control technique has been applied in sustainable transportation systems [94], optimal information diffusion on social networks [95, 96], power networks, among many others.

12.3.4 Illustrative Example

FIGURE 12.3: Optimal time car problem.

Consider the car shown in Figure 12.3. The distance traveled by the car at time t is denoted by $x(t)$. The maximum acceleration and deceleration are bounded between M and $-M$. The dynamics of the system can be represented as

$$\ddot{x}(t) = u(t)$$

or in state variable for

$$\begin{cases} \dot{x}_1(t) = x_2(t) \\ \dot{x}_2(t) = u(t), \end{cases} \quad (12.23)$$

where $x_1(t)$ and $x_2(t)$ are state variables representing the position and velocity of the car, respectively. $u(t)$ represents the control action (acceleration or deceleration) of the car.

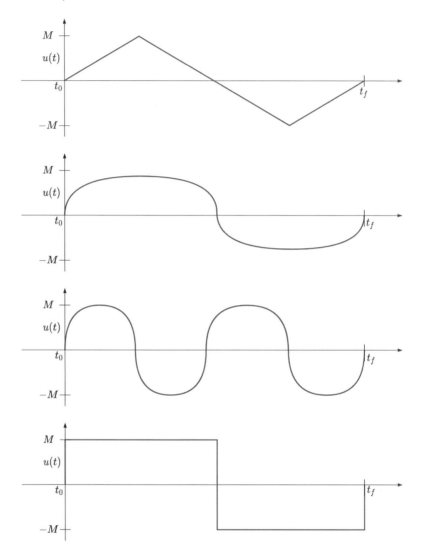

FIGURE 12.4: Sample trajectories.

Figure 12.4 shows possible control trajectories of the car. The question arises, which control trajectory (acceleration and deceleration profiles) does one choose? The answer lies in what is your control objective you selected. Next, we describe a few standard control objectives and their corresponding performance measure.

Objective 1: *To drive the car from point A to point B in the minimum amount of time.* This is a minimum-time problem and its performance measure can be described as

$$J = \int_0^{t_f} dt,$$

i.e. $J = t_f$, where t_f is the time car reaches point B.

Objective 2: *To drive the car from point A to point B with minimum fuel expenditure.* This is a minimum-control problem and its performance measure can be described as

$$J = \int_0^{t_f} u^2(t)dt,$$

where we assume that acceleration and deceleration are directly proportional to the amount of fuel consumed by the car.

Objective 3: *To drive the car from point A to point B in the minimum time while minimizing the fuel expenditure.* This is a mixed-control problem and its performance measure can be described as

$$J = \int_0^{t_f} [Ru^2(t) + \lambda]dt,$$

where λ is the weight assigned to minimum time component of the performance measure, whereas R is the weight assigned to the minimum control component of the performance measure.

More formally, consider an optimal control problem to drive the system described by

$$\begin{cases} \dot{x}_1(t) = x_2(t) \\ \dot{x}_2(t) = u(t) \end{cases} \tag{12.24}$$

with control constraints as $-M < u(t) < M$, in minimum time from point A to point B, from rest to rest, i.e. $x_2(0) = 0$ and $x_2(t_f) = 0$.

It can be derived that the optimal trajectories $u^*(t)$, $x_1^*(t)$, and $x_2^*(t)$ are as shown in Figure 12.5. On a closer look, notice that the $u^*(t)$ trajectory means that to drive the car to the destination in minimum time, one needs to apply maximum throttle ($u(t) = M$) during the first half of the journey and then apply the maximum amount of brakes ($u(t) = -M$) during the last half of the journey. Note that this driving strategy is time optimal but not control optimal, i.e. it will consume a significantly larger amount of fuel!

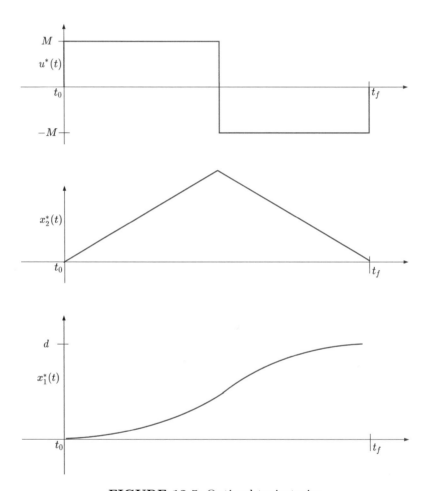

FIGURE 12.5: Optimal trajectories.

12.4 Exercises

1. The second order mechanical system has the state space form as

$$\dot{x}(t) = \begin{bmatrix} 0 & 1 \\ 1 & 1 \end{bmatrix} x(t) + \begin{bmatrix} 100 \\ -50 \end{bmatrix} u(t)$$

$$y(t) = \begin{bmatrix} 1 & 0 \end{bmatrix} x(t)$$

   a) Check the observability of the system.

   b) Design the Luenberger observer for the system so that the closed loop poles are placed at -1 and -2.

   c) Write the observer equation using obtained observer gains in the previous part.

2. A system dynamics is given as

$$\dot{x}(t) = \begin{bmatrix} -4 & 0 \\ 1 & -1 \end{bmatrix} x(t) + \begin{bmatrix} 1 \\ 0 \end{bmatrix} u(t)$$

$$y(t) = \begin{bmatrix} 0 & 1 \end{bmatrix} x(t) + \begin{bmatrix} 0 \end{bmatrix} u(t)$$

   In order to place the closed loop poles at $S = -1 \pm 3j$ find the required state variable feedback assuming that complete state vector is available.

3. Consider the system

$$\dot{x}(t) = \begin{bmatrix} -6 & 2 & 0 \\ 4 & 0 & 7 \\ -10 & 1 & 11 \end{bmatrix} x(t) + \begin{bmatrix} 5 \\ 0 \\ 1 \end{bmatrix} u(t)$$

$$y(t) = \begin{bmatrix} 1 & 2 & 1 \end{bmatrix} x(t)$$

   Using ctrb and obsv functions in MATLAB, show that the system is both controllable and observable.

4. Find a feedback gain matrix K so that the closed poles of the system described by

$$\dot{x}(t) = \begin{bmatrix} 0 & 1 \\ -1 & 2 \end{bmatrix} x(t) + \begin{bmatrix} 1 \\ 1 \end{bmatrix} u(t)$$

$$y(t) = \begin{bmatrix} 1 \\ -1 \end{bmatrix} x(t)$$

are located at $S_1 = -1$ and $S_2 = -2$. Use state feedback control law as $u(t) = -Kx(t)$.

5. The dynamics of a system are given as:

$$\dot{x}_1(t) = x_2(t)$$
$$\dot{x}_2(t) = -x_2(t) + u(t)$$

and the cost function to be minimized is

$$J = \frac{1}{2} \int_0^2 u^2(t)dt,$$

Optimal feedback solution is to be found by using Pontryagin's Minimization Principle. Admissible states and controls are not bounded. $X(0) = [0 \ 0]'$ and $X(2) = [5 \ 2]'$.

a) Find the necessary conditions that must be satisfied. (Obtain state equations and co-state equations).

b) Try to solve analytically using the necessary conditions and boundary values.

c) Develop a MATLAB code to solve the new ODE system symbolically using *syms* and *dsolve* functions.

d) Compare the results obtained in part (b) and (c) by plotting relevant state trajectories.

6. Given the system dynamics

$$\dot{x}_1(t) = x_2(t)$$
$$\dot{x}_2(t) = -x_1(t) + x_2(t) + u(t)$$

and the cost function to be minimized as

$$J = \int_0^T \frac{1}{2} [\beta_1 x_1^2(t) + \beta_2 x_2^2(t) + \beta_3 u^2(t)]dt; \qquad \beta_1, \beta_2, \beta_3 > 0.$$

Using the HJB approach, find the optimal control $U^*(t)$ expressed as a function of $X(t)$, t, and J_X^*.

13

Information Spread and Control

> *He who influences the thought of his times, influences all the times that follow. He has made his impress on eternity.*
>
> Hypatia

In the previous two chapters, an intuitive and fundamental overview of control systems was presented. Here, the same principles will be discussed, but with the goal of applying these principles to information spread. Because there is no one right method to use for any given information spread situation, a certain amount of intuition and creativity are required to design control schemes. To this end, socio-technical systems (of which social media systems are included) are presented as an intersection of social and technical subsystems along with external subsystem influences that must be considered when designing control methods for information spread.

By the end of this chapter, the reader should have a better grasp on identifying possible control methods for social media information spread applications, as well as potential control actions to drive the systems to the desired evolution.

13.1 Controlling Socio-technical Systems

There are three main concerns in socio-technical systems when applied to social media: technical subsystems, social subsystems, and external subsystems.

- **Technical subsystems** include processes, tasks, required technology, structures, and technical specifications necessary to access and use social media. For example, many people use smartphones to create content for social media, and therefore the technical specifications of a smartphone may be a technical subsystem. The internet, of course, is a required technology for social media access.

- **Social subsystems** are concerned with human factors, such as our physical form, skills, values, social structures, psychology, and relationships. Would a miniaturized desktop computer with communication capabilities be as useful or practical as a smartphone? Certainly not. Typing on a tiny keyboard is comical at best, not to mention wildly unusable. Though compliant with the technical needs, such a device would not be designed with people in mind due to our lack of skill to utilize or carry it. A human-sounding voice for digital assistants is another good example, as it helps users feel more connected to the object.

- **External subsystems** are outside influences, such as laws, regulations, business and profit decisions, politics, and more. Perhaps there are energy efficiency standards that must be met or laws requiring the smartphone developers to allow consumers to choose their own browser (and not be locked into one bundled with the phone). Designers are constrained by these external influences when developing the technical and social aspects of their devices or applications.

The simplest way to think of socio-technical systems is as a Venn diagram of social and technical subsystems with external influences, as shown in Figure 13.1. In this high-level interpretation, social and technical subsystems intersect as socio-technical systems. Note that external influences can affect one or both of the subsystems. Perhaps a law prohibits designing a social media site without notifying the user how their data will be used, while shareholders would like to monetize user participation. Maybe there's politicized public anger about how the same social media site sells data or is used to spread misinformation without moderation.

In a more dynamic visualization, subset domains of social and technical systems can be viewed discretely. Notice in Figure 13.2 how in each domain interacts with and influences one another. Social domains such as people, culture, societal and individual goals, and societal norms display a coupled bidirectional interaction with not only each other but also with technical domains such as infrastructure and technology.

With this in mind, we can see how special considerations must be met when designing systems for information spread (such as online social media) and in attempts to control these systems. The reason for this is simple: they are incredibly complicated and non-deterministic. Not only are there technical elements to account for (which are largely consistent and understood), but also social elements. These social elements can vary considerably from group to group or person to person. Everyone has different values, self-identity, abilities, opinions, and personalities. A wealthy and elderly individual will see the world differently than a poor college student and interact with social media and the internet differently. Constantly shifting external factors share similar challenges.

Often, one's goal is to understand the relevant social structures, campaigns, leaders, cultural positives and negatives, and ultimately influence them via control (in a constructive or destructive manner). To this end, one needs to know how the social aspects of the system fundamentally and mathematically

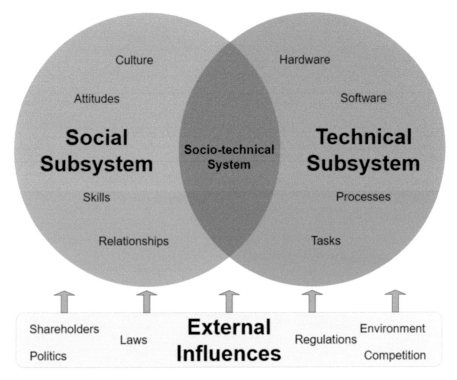

FIGURE 13.1: Social and technical subsystems overlap as socio-technical systems.

behave in order to control it. This behavior is captured in the system models. To apply control theory, a reasonable model of the system must exist or be established. Existing models are very good at explaining socio-technical systems to an extent, but context-aware (or scenario-based models) are sometimes needed. You can see several examples of context-aware models in previous chapters to handle specific instances of social media interactions.

Though some models are intuitive, they must still be run, tested, and validated before control is designed. Since we cannot control based solely on intuition, a model is required. Additionally, the application of well-established engineering control schemes requires closed-form expressions. As such, we must formulate macroscopic mathematical models in the form of equations to apply these control methods. For this reason, the previous chapters focused significant attention on existing closed-form models and their modifications for using them to social media information spread. It is essential to have these closed-form expression models to apply existing control techniques.

For our purposes, macroscopic dynamical models are beneficial for applying these accepted control techniques. Because each individual in a group will have varying whims and tendencies on social media networks, they cannot be

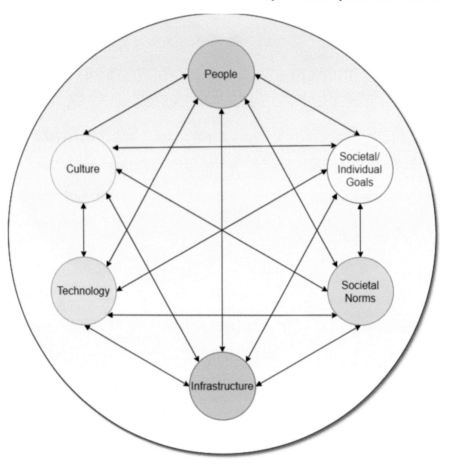

FIGURE 13.2: Social and technical domains interact as part of socio-technical systems.

accounted for individually. As such, they must be looked at in aggregate. In other words, the people must be treated as one group (or at least as a manageable number of subgroups). To do this, we base assumed behavior around "typical" behavior, for example, users reposting stories they are interested in. Obviously, by generalizing and aggregating a population, unavoidable error will be introduced, but the same fundamental control principles will still apply once a good macroscopic model is formulated.

13.2 The Control Action and Social Media Systems

As with standard control application, there are two types of control actions commonly used: open-loop and closed-loop. As before, for open-loop control systems, the control action is independent of the process variable. Likewise, for closed-loop control systems, the control action is dependent on the process variable.

But how does one apply a control action to a social media system or information spread? Let's take some examples from our earlier case study. Consider the previously proposed social campaign model. The process flow diagram is shown in Figure 13.3. Recall the mathematical model developed for the context-aware social media marketing model:

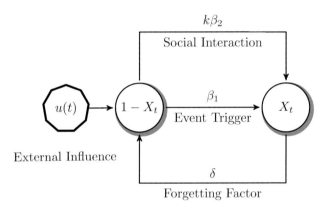

FIGURE 13.3: Flow diagram for event-triggered social media chatter model.

$$dX_t = \beta_1 u(t)[1 - X_t] + k\beta_2[1 - X_t]X_t - \delta X_t. \tag{13.1}$$

The control input (trigger) here is represented by $u(t)$ and is an external influence driving the system evolution. In other words, each major event sparks social media chatter about the topic. Specifically, the control trigger is acting upon those unaware of the event, hence the $u(t)[1 - X_t]$ term. Indirectly, via online social interaction, this control action then influences the rest of the system. Remember to always identify what terms are interacting with each other. Do all interactions make intuitive sense?

Naturally, the same principles can be similarly applied to other situations for a more deliberate control effort. For example, a music artist might slowly release songs to the public in order to build and maintain an online media presence to better sell an album. A tech company can tease small specifications of their latest gadget (through rumors or interviews) to keep it socially relevant and discussed on social media. If any of these scenario-specific models have special limitations or differences, they must be taken into account at the modeling level. Advertising the latest book from a prominent author may have more impact before release than after being consumed by the public. Once the book is released, any rumors concerning plot details will be resolved, and the interest will naturally end.

13.3 Optimal Control and Social Media

Optimal control methods are particularly well suited for social media-based applications. As with most forms of controlling a system that evolves along a natural trajectory, resources must be spent strategically and compromises must be made. The resources used to control social media systems can typically be reduced to capital expenditure (money spent). However, sometimes resources take other forms such as a grassroots effort for a cause, where emotional appeals evoke more attention than neutral ones. In our case, optimal control methods used on social media systems will focus on capital resources since it is quantifiable and easily visualized. Money is spent as a resource to run internet banner ads, sponsored posts, send samples to review bloggers, sponsor YouTube channels, and hire social media marketers.

The performance measure (or cost function) **J** expresses the strategy by which these resources will be spent. While everyone would surely like their influence over a social media group (be it for a political campaign, product, or simple public announcement) to be swift, precise, and inexpensive, reality forces us to choose some elements we favor over others. Would you rather quickly educate the public about how to protect themselves during an emergency or would you rather give slower, more detailed guidelines? Is it better to run a set of political campaign ads over inexpensive online ad space for a longer period of time or to flood the public with them close to election day at a higher cost? There is no right or wrong answer, only context-dependent answers. Examples of optimal control in social media will be further explored as concrete examples in Chapters 14 and 15.

13.4   Exercises

1. Explain what is meant by a "socio-technical system". How is it different from conventional engineering systems?

2. What are some of the challenges to be overcome when trying to apply control schemes to socio-technical systems?

3. Choose an example of a socio-technical system not already discussed in this chapter. List some of the elements that make up its social and technical subsystems. What kinds of specific external influences must be considered when dealing with the system?

4. Some examples of control actions for social media systems were presented in the text. Give another example of a control action that can be applied over social media groups. Are they practical to implement? Explain your reasoning.

14

Control Application 1: Advertisements and Social Crazes

" Advertisers constantly invent cures to which there is no disease. "

Leonardo da Vinci

This chapter is the first of two chapters that expand upon the optimal control principles learned in previous sections by formulating a practical social media information spread problem and step-by-step work toward an optimal control solution to that problem. The first control application is concerned with social marketing campaigns in an attempt to maintain social media interest in a product, service, or event. The example begins with a description of the problem, including some assumptions, goals, and qualitative parameters that must be addressed. Next, the problem is formulated based on the description and translated into a mathematical model (the proposed event-triggered model for social media chatter). Optimal control techniques previously discussed in the text including defining the model to be utilized, an objective function, and an optimal control strategy are integrated into the problem as well. The results are calculated and the solution is simulated and compared with a non-optimal intermittent control scheme. A brief discussion of the control application in light of the example results concludes the chapter.

14.1 Scenario Description

The spread of a message via advertisements is critical for most businesses and organizations that rely on popular participation to succeed. If the population is unaware of a product or service, it will not be purchased or utilized. As such, the goal of advertising is to spread said message as effectively as possible within the bounds of the spreading organization's time and resources. How does one group advertise to another? By hiring individuals as information spreaders who then spread that message in the form of an ad on television, social media, or billboards. In addition, recent years have seen paid ads in the form of search engine and online product search priority for ad purchasers. Without hired advertisements, the single company spreader would simply never gain enough traction in the public eye to spread to any significant amount of customers.

Ideally, an organization could simply pour an infinite amount of control via ad costs and repetition to the public which would absorb the information and never forget it. Eventually, everyone would know about the product or service and its benefits, giving the best opportunity for advertisers to gain as many customers as possible. Unfortunately, this ideal situation is rarely if ever realistic. Resources are limited by the pocket depth of the advertisers as well as other practical constraints. Additionally, customers are likely to forget an advertisement's message over time and requiring additional ads to be spread to maintain attention.

But what happens when advertisements cease? The message does not simply disappear from the public consciousness. In the absence of active advertising, especially in a digital social media world, talk of the product or service will endure, perhaps even thrive and grow into what is known as a "social craze". If a product, service, social movement, or meme could sustain itself without active advertising, despite decay, it would ultimately prove beneficial and cost effective for the original advertiser.

14.2 Problem Formulation

The control goal for this scenario is to create a social media craze or chatter at a certain desired level using active advertising. Ideally, the advertisers who wish to start this craze will spend only as much as is required to make the information spread beyond its natural chatter levels for a time, without requiring additional advertisement resources. At first glance, the problem appears to closely mirror the Ignorant-Spreader-Recovered (ISR) model presented earlier. Alternatively, a spreading rate model with the addition of a decreasing "forgetting" factor (as product excitement dwindles for a particular model or iteration) can be developed. Both proposed modelings of the advertisement spread account for the main elements of the problem: the active spread of the information and a natural tendency of interest in the product to diminish. However, the proposed event-triggered model for social media crazes or chatter fits the scenario nicely.

Recall that the social craze or social media chatter model from Equation 9.8 was presented as

$$dX_t = \beta_1 u(t)[1 - X_t] + \beta_2[1 - X_t]X_t - \delta X_t,$$

where β_1 is the social marketing campaign constant, β_2 represents the social craze constant, δ is the decay constant, and $u(t)$ is the control, expressed through active advertising initiatives.

Recall also that our goal is to pass the advertising socio-equilibrium threshold and then to sustain the social craze or chatter at a certain level in the presence of active control. Therefore, active control must be maintained until the chatter has reached the desired level. Once the control is stopped, the activity climbs down to the equilibrium point defined as:

$$X_{eqb} = 1 - \frac{\delta}{\beta_2}.$$

Because we wish to minimize the required expenses of the advertisers and accomplish the goal of achieving social craze status as quickly as possible over the course of the campaign timeline, we take the cost function to be:

$$J = \int_0^{t_f} \left[u^2(t) + (x - x_d)^2 + \lambda \right] dt, \tag{14.1}$$

where λ is weight placed on the time and x_d is the desired amount of activity (information spreaders). Once, there is no further need to maintain the desired level of activity, the control action $u(t)$ will cease and the activity level will be back at X_{eqb}.

This scenario is a prime candidate for an optimal control strategy. There are any number of possible ways to achieve the objective. It can be weighted to minimize costs, progress as fast as possible, use strict control for precision, or any weighted combination of objective elements. Other scenarios similarly utilize optimal control principles to find the best solution for a given need [97]. The Hamilton-Jacobi-Bellman (HJB) equation, as the basis of optimal control, will be utilized to achieve an optimal control strategy, $u^*(t)$, representing how much advertising should be put out on social media over time. In most cases, no analytical solution of a PDE is attainable and even a numerical solution is challenging. To overcome this, the Pontryagin's minimization principle will be applied to the problem, which yields a relatively simple numeric solution.

14.3 Dynamic Programming Approach

The Hamiltonian is calculated, as follows:

$$\mathcal{H} = Ru^2(t) + (x(t) - x_d)^2 + \lambda + J_x[\beta_1(1 - x(t))u(t) \\ + \beta_2(1 - x(t))x(t) - \delta x(t)]. \tag{14.2}$$

By differentiating the Hamiltonian, with respect to the control $u(t)$ and setting the result equal to zero, the optimal control action is found by solving for $u^*(t)$:

$$u^*(t) = \frac{J_x \beta_1 (1 - x(t))}{2R}. \tag{14.3}$$

Finally, the resulting $u^*(t)$ control action is used to determine the HJB partial differential equation,

$$J_t^* + \frac{J_x^2 \beta_1^2 (1 - x(t))^2}{4R^2} + J_x x(t)(\beta_2 x(t) - \delta) = 0. \tag{14.4}$$

Because there is no clear analytic solution from the resulting HJB partial differential equation, another method is required.

14.4 Pontryagin's Approach

The Pontryagin's minimization principle is applied to the same dynamics and cost function. The Hamiltonian is calculated using the new method:

$$\mathcal{H} = g + p^T [a] = Ru^2(t) + M(x(t) - x_d)^2 + \lambda + p[\beta_1 u(t)(1 - x) \\ + \beta_2(1 - x(t))x(t) - \delta x(t)], \tag{14.5}$$

where R and M are weight constants for the control and tracking elements of the cost function, respectively. From the Hamiltonian, the state and co-state equations are determined as follows:

$$\begin{cases} \dot{x}(t) = \dfrac{\delta \mathcal{H}}{\delta p} = \beta_1 u(t)(1 - x(t)) + \beta_2(1 - x(t))x(t) - \delta x(t) \\[2mm] \dot{p}(t) = -\dfrac{\delta \mathcal{H}}{\delta x} = 2M(x_d - x(t)) + p(\beta_1 u(t) + \beta_2(2x(t) - 1) + \delta). \end{cases} \tag{14.6}$$

By differentiating the Hamiltonian with respect to the control $u(t)$ and setting the result equal to zero, the optimal control action is found as follows:

$$\frac{\delta \mathcal{H}}{\delta u(t)} = 2Ru(t) + p\beta_1(1 - x(t)) = 0, \tag{14.7}$$

$$u^*(t) = \frac{-p^* \beta_1(1 - x^*(t))}{2R}. \tag{14.8}$$

Using the calculated $u^*(t)$, along with the state and co-state equations, the necessary conditions can be expressed analytically as

$$\begin{cases} \dot{x}^*(t) = \dfrac{-p^*(t)\beta_1^2}{2}(x^{*2}(t) - 2x^*(t) + 1) + \beta_2(1 - x^*(t))x^*(t) - \delta x^*(t) \\[2mm] \dot{p}^*(t) = \dfrac{p^{*2}(t)\beta_1^2}{2}(x^*(t) - 1) - 2p^*(t)\beta_2 x^{*2}(t) - p^{*2}(t)\lambda + 2x_d. \end{cases} \tag{14.9}$$

14.5   Numerical Solution and Discussion

Using `MATLAB`, the boundary value problem is simulated with the *bvp4c* function. The resulting plots of normalized population versus time for identical parameters are shown in Figure 14.1.

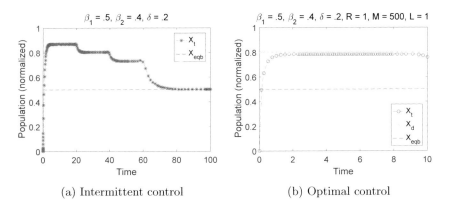

(a) Intermittent control   (b) Optimal control

FIGURE 14.1: Comparing intermittent and optimal control for an advertisement or social craze.

Notice that by using intermittent control, an initially strong amount of control is exercised on advertising over social media to pass the socio-equilibrium threshold X_{eqb} and raise awareness of the product, service, or event. The active advertising control is slowly reduced at intervals to maintain higher than normal social media chatter until control is halted (resources are no longer being spent to advertise) and general interest stabilizes at the socio-equilibrium threshold. However, by applying optimal control methods, social media chatter quickly reaches the desired level and is steadily maintained above the desired natural social media chatter level.

Suppose the given example scenario represents a minor business advertiser whose advertisements only influence a small portion of those who view it. However, the social media group to which the advertisement is targeted is highly interactive and connected. If someone likes the product, they will be very likely to share that information with their friends. It is easy to observe the need for significant control initially, as the social media advertisement system is sustained by the active advertisements. As the system progresses over time, less active advertisement control is needed as more social media "buzz" is accruing and spreading the advertisement information. Eventually, no active control (and hence, advertisement effort and cost) is required to sustain the social media information. It becomes a "craze" and is discussed, re-tweeted, and popular for a time after there is no active control.

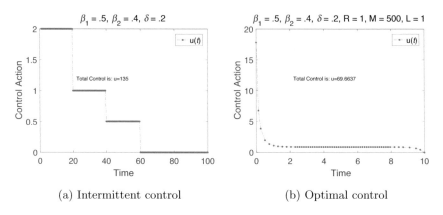

(a) Intermittent control   (b) Optimal control

FIGURE 14.2: Control action trajectories for intermittent and optimal control for an advertisement or social craze.

The trajectory of the control actions for intermittent and optimal control methods can also be compared, as shown in Figure 14.2. Using intermittent control, the advertiser must select specific points in time at which to release a new ad or social media post to maintain popular interest in the subject of the advertisement. With each of these advertisement control actions, social media chatter is kept from degrading completely, but is still driven to its equilibrium threshold when not maintained. The problem with this method, however, is that the advertisements are inefficient, costly, and must often be micromanaged.

By following the optimal control action trajectory, advertising and resource usage is continuous and is smoothly applied in order to reach the desired interest levels. Comparing the total control exercised, the optimal control method achieves the desired objective much more quickly than intermittent control and at significantly less resource cost. In this specific example simulation, control is heavily used at the start of the advertising campaign and very little advertising is required as time progressed.

Through the model equations and simulations, it is clear that the advertiser can have a significant sway over people through its advertisement (control) actions, as denoted by the β_1 parameter. Additionally, the speed at which the advertisement awareness and spreading reaches the desired social media craze threshold can be influenced. The weight factors in the objective (or cost) function can be adjusted to meet the needs and goals of the advertiser, whether it be the minimizing of resources, time spend, or distance from a desired threshold level.

The impact of the decay factor also cannot be understated. Even with otherwise similar parameters, the decay constant, δ, has a strong effect on the person-to-person social media chatter interactions. In other words, a popular advertiser is advertising to a receptive group that will readily post and share the product, social cause, etc. However, this group can grow disinterested or bored quickly with a high decay constant. As a result, significantly more control is required when attempting to form a social media craze out of little or no initial group interest.

Using the above discussion, it is not difficult to formulate marketing strategies of several companies, that can be similarly modeled and simulated by changing the parameters. Apple, for example, has a strong enough brand following that it needs very little active advertisement to create a group obsession over the latest device. In contrast, governmental health organizations may pour massive amounts of advertising into changing public opinion over bad health habits and achieve very slow progress. As such, achieving optimal spending of resources to gain online social media awareness, discussion, and sharing is especially critical for unpopular or uninteresting messages where a natural socio-equilibrium threshold is comparatively low.

```
Code 14.1: Optimal Control of Social Craze/Advertisement
1  function Social_Craze_BVP_SA
2  %BVP for of optimal control of a social craze.
3  % Begin with an initial solution guess
4  solution_initial = bvpinit(linspace(0,10,40), [0
      0]);
5  optional = bvpset('Stats','on','RelTol',1e-1);
6  global R; % control weight
7  global M; % (x1 - xd)^2 weight
8  global L; % time weight lambda in the cost
      function
9  global B1; % beta_1 = spreading rate due to
      advertisement
10 global B2; % beta_2 = spreading rate due to social
      media
11 global d; % delta = rate of decay due to
      disinterest
12 global x_d; %desired level of activity
13 global x_eqb; %equilibrium value
14 global tf; % final time
15
16 R = 1;
17 M = 500;
18 L = 1;
19 B1 = 0.5;
20 B2 = 0.4;
21 d = 0.2;
22 x_d = .8;
23 x_eqb= 1 - d/B2;
```

```
24  tf = 100;
25
26  solution = bvp4c(@BVP_ode, @BVP_bc,
        solution_initial, optional);
27  t = solution.x;
28  y = solution.y;
29
30  % u(t) is calculated using the states and costates
        (x1 and p1)
31  ut = -(y(2,:)*B1.*(1-y(1,:)))/2*R;
32
33  %Calculate total control
34  u_total = trapz(ut)
35  n = length(t);
36
37  % Calculate the cost
38  J = tf*((M*(y(1,:)-x_d)'*(y(1,:)-x_d)) + R*(ut*ut
        ') + L)/n;
39  figure(1);
40  plot(t, y(1,:)','--mo');hold on;
41  plot(t,ones(size(t))*x_d,'-gx');
42  plot(t,ones(size(t))*x_eqb,'b--'); % plot
        equilibirum line
43  title({'\beta_1 = .5, \beta_2 = .4, \delta = .2, R
        = 1, M = 500, L = 1'}, 'fontweight', 'normal'
        )
44  xlabel('Time');
45  ylabel('Population (normalized)');
46  legend({'X_t','X_d','X_{eqb}'}, 'Location', '
        southeast');
47  %xlim([0 10])
48  ylim([0 1])
49  ax = gca;
50  ax.FontSize = 16;
51  hold off;
52
53  figure(2);
54  plot(t,ut','-.r.','MarkerSize',10)
55  s = strcat('Total Control is: u=',num2str(u_total)
        );
56  text(2.2,12,s);
57  title({'\beta_1 = .5, \beta_2 = .4, \delta = .2, R
        = 1, M = 500, L = 1'}, 'fontweight', 'normal'
        )
```

```matlab
xlabel('Time');
ylabel('Control Action');
legend('u(t)');
ax = gca;
ax.FontSize = 16;
hold off;
%-------------------------------------------------
% ODE's for states and costates
function dydt = BVP_ode(t,y)
x1 = y(1);
p1 = y(2);
u = -(y(2)*B1.*(1-y(1)))/(2*R);
dydt = [B1*(1-x1).*u + B2*(1-x1)*x1' - d*x1
        2*M*(x_d-x1) + p1*(B1*u + B2*(2*x1 -1) + d
          )];
% -------------------------------------------------
% Nested function to do time varying control
    action
%  function u = ctrl(t)
%     if t< 20
%         u = 0;
%     elseif 20<=t && t<100
%         u = -(y(2)*B1.*(1-y(1)))/(2*R);
%     else u = 0;
%     end
%  end
end
%-------------------------------------------------
% The boundary conditions:
% x1(0) = 0, tf = 10, p1(tf) = x_d;
function res = BVP_bc(ya,yb)
res = [ ya(1) - 0        % starting percent of
    spreaders
          yb(2) - x_d]; % desired tracking level
end
end
```

15

Control Application 2: Stopping a Fake News Outbreak

Nothing can now be believed which is seen in a newspaper. Truth itself becomes suspicious by being put into that polluted vehicle. The real extent of this state of misinformation is known only to those who are in situations to confront facts within their knowledge with the lies of the day.

Thomas Jefferson, *The Writings of Thomas Jefferson*, 1854

Continuing along the same methodology presented in the previous control application example, this chapter also formulates a practical social media information spread scenario and proceeds to provide a step-by-step progression toward an optimal control solution to the problem. This second control application example revolves around the mitigation of a fake news or misinformation outbreak over social media. Like the previous scenario, it begins with a description of the problem of stopping a fake news outbreak along with relevant qualitative parameters, goals, and assumptions. The problem is then mathematically formulated through a combination of an ISR-based model for social media information spread and optimal control methods. From a control perspective, the objective function and an optimal control strategy are defined. The calculated results to the control application problem are simulated side-by-side with a simple intermittent control method for the sake of comparison. The relevance of the control application results is discussed briefly at the end of the chapter.

15.1   Scenario Description

Consider a scenario similar to China's Salt Panic in which false news is being spread along not only rumor channels, but also trusted internet news sources. The government is aware of the threat that the news may pose to the community and must take steps to prevent it from becoming an information epidemic, as it may cause mass public panic. The fake news of this type is traditionally fast spreading and the damage from it will be done in short order if the information is not quelled. Obviously, this potential information epidemic has already been spreading around the population before the government became aware of its growing popularity.

Luckily, the government has access to modern emergency alert information and official and direct news distribution over the internet. "Tweets", live YouTube press conferences, and cell phone alerts are all options of quick communication to help stifle the false news. Note that in many ways, it does not matter if the news is real or fake or if the government wishes to spread or diminish the news. The core formulation and strategy of control here are analogous.

15.2   Problem Formulation

The goal of the control action is to ultimately prevent the fake news from taking on a substantial life of its own within the public's social media networks. Recall the concept of herd immunity, where if a sufficient percentage of the population is immunized from the fake news, it will never be able to "take off" into becoming an information epidemic. The idea of a control condition for regulating a desired property at a critical point is a powerful tool in dealing with problems such as this [52]. One strategy by which to do this is to educate the population before they receive word of the fake news. This method effectively shrinks the pool from which spreaders of fake news can pull (the ignorant individuals) such that they are unable to create as many future spreaders of the fake news because there are simply not enough members of the ignorant class remaining to convert. With a sufficiently low number of spreaders and, hence, high number of educated ignorants, the fake news information can never reach an epidemic state.

Recall first the modified ISR model for social media interactions. We will first focus attention on the dynamics of the ignorant and spreader classes, rewritten here as:

$$\begin{cases} \dot{x}_1(t) = -\beta x_1(t)x_2(t) - bu(t)x_1(t) \\ \dot{x}_2(t) = \beta x_1(t)x_2(t) - \gamma x_2^2(t), \end{cases} \tag{15.1}$$

where $x_1(t)$ and $x_2(t)$ are ignorants and spreaders of the fake news, respectively, and the term $-bu(t)x_1(t)$ is the attempt to reduce the number of ignorant individuals by educating them on the false nature of the news. In epidemiology this is often done through immunizations. Also recall from herd immunity theory, that the basic reproductive number R_0 is essentially the ratio of the spreading rate and stifling rate and that for no information epidemic to occur, we must satisfy

$$R_0 = \frac{\beta}{\gamma} < 1. \tag{15.2}$$

Notice that in the above equation parameters β and γ are the properties of an information spread system and the type of news/rumor in consideration. We have little control to adjust these parameters. So how do we bring down the R_0 below 1? The answer is through immunization (educating the population with correct information). Let p be the percentage of the population that is immunized (through education in this case) to stop the fake news epidemic. This immunization reduces the population susceptible to "infection" with fake news. Hence the resulting effective reproduction number becomes $R_0(1 - p)$. This means that we now require $R_0(1 - p) < 1$, or

$$p > 1 - \frac{1}{R_0}.$$

Now using $R_0 = \frac{\beta}{\gamma}$, we get

$$p > 1 - \frac{\gamma}{\beta}.$$

Since the percent recovered population in the ISR model is equal to $1 - I - S$, which is also equal to the desired percentage of group education against the fake news, the equation can be expressed as

$$I + S < \frac{\gamma}{\beta}.$$

Or,

$$x_1(t) + x_2(t) < \frac{\gamma}{\beta}, \tag{15.3}$$

which serves as the control objective in this case. As discussed previously, many control approaches can be adopted to achieve the desired control objectives such as hit and trial, feedback linearization control of non-linear systems, and optimal control. For this example, we adopt an optimal control approach, the most suitable approach for problems where expenditures associated with control actions (i.e. money) pose a practical design constraint.

Design of Cost Function: The desired objective can be achieved using various cost functions. Let us first consider a cost function where the desired immunization has to be achieved within a given time frame t_f. In this case, the cost function will take a shape similar to:

$$J = \left\| x_1(t_f) + x_2(t_f) - \frac{\gamma}{\beta} \right\|^2. \tag{15.4}$$

A more general cost function would try to balance control action costs and time consumed while trying to achieve the control objective:

$$J = \int_0^{t_f} \left(Ru^2(t) + Q[x_1(t) + x_2(t)]^2 + \lambda \right) dt, \tag{15.5}$$

where $u(t)$ is the control exercised by public social media posts, announcements, advertisements, text alerts, etc. The constants R, and Q are adjustment weight factors to put stronger or weaker emphasis on each element of the cost function. λ is the weight on time elapsed during the process, which serves as the control objective in this case. The Pontryagin minimization method will be used to strike a balance between amount of control exerted, the sum of the ignorant and spreading individuals, and time.

15.3 Pontryagin's Approach

Following the Pontryagin's minimization principle, the Hamiltonian is calculated, as follows:

$$\mathcal{H} = (Ru^2(t) + Q(x_1(t) + x_2(t))^2 + \lambda) + p_1[-\beta x_1(t)x_2(t) - bu(t)x_1(t)]$$
$$+ p_2[\beta x_1(t)x_2(t) - \gamma x_2^2(t)]. \tag{15.6}$$

The state and co-state equations are calculated from the dynamics, the cost function, and the Hamiltonian:

$$\begin{cases} \dot{x}_1(t) = \dfrac{\delta \mathcal{H}}{\delta p_1} = -\beta x_1(t)x_2(t) - bu(t)x_1(t) \\[2mm] \dot{x}_2(t) = \dfrac{\delta \mathcal{H}}{\delta p_2} = \beta x_1(t)x_2(t) - \gamma x_2^2(t) \\[2mm] \dot{p}_1(t) = -\dfrac{\delta \mathcal{H}}{\delta x_1(t)} = -[2Q(x_1(t) + x_2(t)) - p_1\beta x_2(t) - bu(t)p_1 + p_2\beta x_2(t)] \\[2mm] \dot{p}_2(t) = -\dfrac{\delta \mathcal{H}}{\delta x_1(t)} = -[2Q(x_1(t) + x_2(t)) - p_1\beta x_1(t) + p_2\beta x_1(t) - 2\gamma p_2 x_2(t)]. \end{cases} \tag{15.7}$$

By differentiating the Hamiltonian with respect to the control $u(t)$ and setting the result equal to zero, the optimal control action can be found as follows:

$$\frac{\delta \mathcal{H}}{\delta u(t)} = 2Ru^*(t) - p_1^* bx_1^*(t) = 0, \tag{15.8}$$

and

$$u^*(t) = \frac{-p_1^* bx_1^*(t)}{2R}. \tag{15.9}$$

15.4   Numerical Solution and Discussion

Using `MATLAB`, the boundary value problem is solved with the `bvp4c` function. Initial conditions $x_1(0)$ and $x_2(0)$ are the initial normalized populations of those ignorant of the fake news or misinformation and those already actively spreading it, respectively. Values for $x_1(t_f)$ and $x_2(t_f)$ will tend towards zero, as the entire population is either educated or stifled from spreading the misinformation. The sum of the two groups must be below the information epidemic threshold stated in the control objective to prevent a misinformation epidemic. In other words, enough people must be educated (either directly or indirectly through online social interactions) or bored with the misinformation. The resulting plots of states versus time are shown in pairs for ease of visual understanding, the first displaying the states. Each subplot displays the ignorant state $x_1(t)$ and the spreading state $x_2(t)$. Figure 15.1 gives a simulation of a widely believed piece of misinformation over social media.

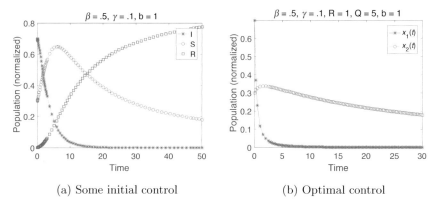

(a) Some initial control   (b) Optimal control

FIGURE 15.1: Comparing random initial control and optimal control to mitigate a misinformation outbreak.

In the first simulated case, is shown in Figure 15.1a. Random initial control is attempted to mitigate and educate those not aware of the misinformation via posts and tweets by authority figures, experts, etc. Unfortunately, the amount of initial control was not sufficient to prevent a misinformation epidemic and cost a considerable amount of resources. Figure 15.1b uses optimal control methods to quickly mitigate the spread of misinformation before it spreads much further. The initial conditions of spreaders spreading fake news and misinformation is already dangerously close to achieving a misinformation epidemic. If such an epidemic were to occur, then nearly everyone at some point would believe the misinformation (however briefly). The longer an uncontrolled information epidemic is allowed to run its course on social media, the more resources and efforts it would require to mitigate it as it quickly grows.

Ideally, as few as possible resources should be consumed in attempting to mitigate the misinformation outbreak. Consider Figure 15.2, which compares the control objective, control trajectories, and total control of the same simulated scenario.

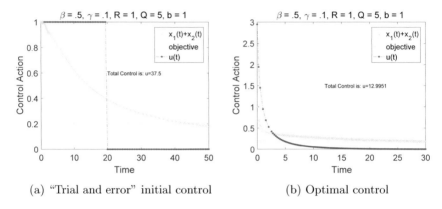

(a) "Trial and error" initial control   (b) Optimal control

FIGURE 15.2: Comparing random initial control and optimal control trajectories to mitigate a misinformation outbreak.

Utilizing an optimal control scheme, the most control is required at the earliest time possible. Spending resources on advertisements, paid social media posts, and public service alert texts would be ineffective if the allowable resources were spent at regular intervals instead of being front weighted and "getting ahead" of the fake news story. Once the total percentage of spreaders begins to decline, then control can be eased considerably. Ultimately, no control is necessary once there are insufficient ignorants left to learn the fake news story or misinformation and the remaining information spreaders will naturally decay with no receptive audience left. Here, the objective $x_1(t) + x_2(t)$ (or the total ignorants and spreaders) must achieve a level below the information epidemic threshold. The control $u(t)$ is used to meet the objective.

Note that the total control is significantly smaller when using the optimal control action in Figure 15.2b over the simple initial control attempt simulated in Figure 15.2a. This results in less resources being spent overall to achieve the desired objective. Additionally, the optimal control method allows for a faster mitigation of the misinformation and prevents a runaway misinformation epidemic. Depending on the weights placed on the objective function constants R, Q, and b the problem solution can be modified to optimize for minimized error in tracking the information epidemic threshold, time to mitigate the misinformation, or total control and resources spent to obtain group immunity to the fake news. A practical balance between the three can also be the objective, depending on the needed requirements.

It is also possible to encourage a fake news or misinformation epidemic, such that most of the population learns the misinformation and actively spreads for a time. The problem would be similarly performed, but with a modified control objective and utilizing the control action to actively advertise false information to the ignorant class. The ignorant class would be similarly "educated" with a strong belief against the truth and faith in the misinformation.

```
Code 15.1: Optimal Control of Containing A Fake News Outbreak
1  function Herd_Immunity_bvp_SA
2  %BVP for optimal control of fake news.
3  % Begin with an initial solution guess
4  solution_initial = bvpinit(linspace(0,30,60), [0 0
       0 0]);
5  optional = bvpset('Stats','on','RelTol',1e-1);
6  global R; % control weight
7  global Q; % (x1 + x2)^2 weight
8  global L; % time weight lamba in the cost function
9  global B; % beta = spreading rate
10 global G; % gamma = stifling rate
11 global b; % weight of control
12 global tf; % final time
13
14 R = 1;
15 Q = 5;
16 L = 1000;
17 B = .5;
18 G = .1;
19 b = 1;
20 tf = 30;
21
22 solution = bvp4c(@BVP_ode, @BVP_bc,
       solution_initial, optional);
23 t = solution.x;
24 y = solution.y;
25 %global p_d;
26 objective = G/B % (x1 + x2) desired objective
27 % u(t) is calculated using the states and costates
       (x1,x2, p1, p2)
28 ut = (y(3,:)*b.*y(1,:))/(2*R);
29 % Calculate total control
30 u_total = trapz(ut)
31 %u_total = trapz(ut)
32 n = length(t);
```

```
33  % Calculate the cost
34  J = tf*(ut*ut'*R +Q*(y(1,:)+y(2,:))*(y(1,:)+y(2,:)
       )' + L)/n;
35  figure(1);
36  %plot(t, y(1:2,:)','-');hold on;
37  plot(t, y(1,:)','-r*');hold on;
38  plot(t, y(2,:)','--mo');hold on;
39  title('\beta = .5, \gamma = .1, R = 1, Q = 5, b =
       1', 'FontWeight','Normal')
40  xlabel('Time');
41  ylabel('Population (normalized)');
42  legend('x_1(t)','x_2(t)');
43  ax = gca;
44  ax.FontSize = 16;
45  hold off;
46
47  figure(2);
48  plot(t, (y(1,:)+y(2,:))','c-x'); hold on; %plot
       ignorant and spreaders
49  plot(t,ones(size(t))*objective,'g--'); % plot
       objective
50  plot(t,ut', '-.r.','MarkerSize',10) % plot control
51  %plot(t, 1-(y(1,:)+y(2,:))','b-o'); hold on; %
       plot recovered
52  s = strcat('Total Control is: u=',num2str(u_total)
       );
53  text(12,1.5,s);
54  title('\beta = .5, \gamma = .1, R = 1, Q = 5, b =
       1', 'FontWeight','Normal')
55  ylabel('Control Action');
56  xlabel('Time');
57  %ylabel('States');
58  legend('x_1(t)+x_2(t)','objective','u(t)');
59  ax = gca;
60  ax.FontSize = 16;
61  hold off;
62  %-------------------------------------------------
63  % ODE's for states and costates
64  function dydt = BVP_ode(t,y)
65  global R;
66  global Q;
67  global B;
68  global G;
69  global b; %weight of control
```

```
70  x1 = y(1);
71  x2 = y(2);
72  p1 = y(3);
73  p2 = y(4);
74  u =  y(3)*b.*y(1)/(2*R);
75  dydt = [-B*x1.*x2-b.*u.*x1
76           B*x1.*x2-G*x2*x2
77          -2*Q*(x1+x2) + p1*B.*x2 + p1*b.*u - p2*B.*
                x2
78          -2*Q*(x1+x2) + p1*B.*x1 - p2*B.*x1 + 2*p2*
                G.*x2];
79  % ------------------------------------------------
80  % The boundary conditions:
81  % x1(0) = 0.7, x2(0) = 0.1, tf = 100, p1(tf) = 0,
       p2(tf) = 0;
82  function res = BVP_bc(ya,yb)
83  res = [ ya(1) - 0.70 %0.90 starting ignorant
84          ya(2) - 0.30 %0.10 starting spreaders
85          yb(3) - 0
86          yb(4) - 0 ];
```

Code 15.2: ''Trial and Error'' Control Containing A Fake News Outbreak

```
1   %Graphs the time evolution of the dynamics
2   clear;
3   to = 0;
4   tf =50;
5   yo = [.7 .3 0];
6   [t y] = ode45('ypISR_sm_control',[to tf],yo);
7   G = .1; %.03;  %stifling constant
8   B = .5;
9   b = 1;
10  objective = .001*G/B; % (x1 + x2) desired
       objective
11
12  figure(1);
13  plot(t,y(:,1),'-.r*',t,y(:,2),'--mo',t,y(:,3),':bs
       ')
14  title('\beta = .5, \gamma = .1, b=1', 'FontWeight'
       ,'Normal')
15  xlabel('Time')
```

```
16  ylabel('Population (normalized)')
17  legend('I','S','R','Location','northeast')
18  ax = gca;
19  ax.FontSize = 16;
20
21  time = 1:.5:50;
22  u = zeros(1,length(time));
23  u(time>=0 & time<=20) = 1;
24  u(time>=20 & time<=100) = 0;
25
26  %Calculate total control
27  u_total = trapz(u)
28
29  figure(2);
30  plot(t, (y(:,1)+y(:,2))','c-x'); hold on; %plot
        ignorant and spreaders
31  plot(t,ones(size(t))*objective*100,'g--'); % plot
        objective
32  plot(time,u','-.r.','MarkerSize',10) % plot
        control
33  %plot(t, 1-(y(1,:)-y(2,:))','b-o'); hold on; %
        plot recovered
34  s = strcat('Total Control is: u=',num2str(u_total)
        );
35  text(20,.6,s);
36  title('\beta = .5, \gamma = .1, R = 1, Q = 5, b =
        1', 'FontWeight','Normal')
37  xlabel('Time');
38  ylabel('Control Action');
39  legend('x_1(t)+x_2(t)','objective','u(t)');
40  ax = gca;
41  ax.FontSize = 16;
42  hold off;
```

Code 15.3: ''Trial and Error'' Control Containing A Fake News
Outbreak

```
1  function ypISR_sm = ypISR_sm_control(t,y)
2  %This function outputs the dynamics to be analyzed
3  %   Detailed explanation goes here
4  G = .1; %.001;  %stifling constant
5  B = .5; %.01;   %spreading constant
6  k = 1; % 1;  %interaction constant
7  b= 1; % control constant
```

```
ut = ctrl(t); % control profile
ypISR_sm(1) =-B*k*y(1)*y(2) -b*ut*y(1) ;
ypISR_sm(2) = B*k*y(1)*y(2)-G*k*y(2)*(y(2));
ypISR_sm(3) = G*k*y(2)*(y(2));
ypISR_sm = [ypISR_sm(1) ypISR_sm(2) ypISR_sm(3)]';


%nested function to do time varying control action
 function ut = ctrl(t)
    if t< 1
        ut = 0;
    elseif 1<=t& t<2
        ut = .1;
    else ut = 0;
    end
 end


end
```

16

Concluding Thoughts

> *The more you know, the more you know you don't know.*

Aristotle

The overriding purpose of this text was to provide an overview of social media information spread theory, modeling, and analysis, as well as provide some example techniques on how to control information spread in various scenarios. It was our goal to allow beginning and advanced readers alike to gain benefit, insight, and inspiration from the topics explored. We hope that readers feel encouraged to engage and communicate their insights into these topics and more, as that is the true value of social media.

This final chapter of the text begins with a brief overview of the topics presented over previous chapters. Information categorization and the importance of studying information spread over social media is discussed. The main social network theory concepts that the reader should take away from reading the text are listed. General modeling and epidemiology-based and social marketing-based specific modeling domains are reviewed. Finally, the control fundamentals and methods touched upon in the text are briefly summarized. For readers that wish to delve deeper into any of the topics explored within the text, some guidance is given as to how to further progress based on specific areas of interest. The chapter concludes with some potential directions and changes that may take place in future research and applications, while reinforcing the lasting value of the concepts presented.

16.1 What Have We Learned?

Over the course of this book many topics were explored. Given the breadth (and sometimes depth) of these topics, it can feel overwhelming at times to recall this text's main takeaways for you, the reader.

First, we attempted to classify information into several overarching categories. While certainly non-exhaustive and somewhat subjective, it should give you some context for categorizing real-world examples you may encounter and how these examples can be approached in the future via models. You also, hopefully, gained insight into why research into information spread is important and how continued research should be encouraged through the use of modern examples, including political campaigning and fake news.

We continued by looking at social networks, in a general sense. We examined and categorized some of the popular online social networks we encounter in our everyday lives. As new social networks develop and others disappear, the list will become outdated, but you should now have the tools to find existing or new ways to categorize them.

Next, we explored the basics of social network theory and learned some of the relationship and structural forms taken by social network groups. Ideally, you should now be able to draw basic sociograms of groups and identify some key relationship and structural types, such as homophily, filter bubbles, reciprocity, and dyadic, triadic, and balanced relationships. Additionally, you should have a sense of how these networks are analyzed, both conceptually and mathematically. Given a network map, you can identify if a network is especially dense or sparse, which ties are strong or weak, and if there is any clustering or polarization. You should be familiar with the concept of small world networks and be able to make a simple adjacency matrix from a sample sociogram.

The section that followed presented you with the concept of models, starting with a high-level overview of general models, presenting some fundamental steps in model development, and then focusing on mathematical models. While making or applying models can be quite challenging, you should be able to apply the examples to begin formulating your own basic models for new examples.

You were presented with several deterministic (epidemiology-based) models on information spread, each keyed for a certain scenario in which they are best applied. Your specific takeaway from this and following sections is tied to your mathematical background, but we encourage you to take advantage of the code provided and explore these models further. At the very least, you should have an understanding of IS, ISI, and ISR models, as they are commonly used in various forms throughout academic literature. You learned that marketing is also a form of information spreading and marketing models can be likewise applied to online information propagation. You should also be aware that these models can be modeled stochastically, though it will get increasingly mathematically complex. A set of case studies demonstrated how to select a case study, collect data, and apply it to a model in the hopes of showing its correlation to reality. Similar methods can be used for other social media trends, events, or social movements.

The final section introduced several fundamentals of control theory and how these methods can be applied both generally and specifically to social media systems. You should now have a better intuitive and technical understanding of how mathematical models can use existing methods (such as engineering control techniques) in order to potentially influence a social media system. Two optimal control examples served to reinforce this idea.

16.2 But Now What?

This text was designed to help introduce you to several topics of interest to the study of social media information spread. Now it's up to you to determine where your areas of specific interest lie and proceed onward. There are many exciting new areas of research with concrete social media and information applications and equally as many resources to help you begin your journey.

Network-based information spreading is a reasonably well-researched field and there is no lack of academic papers, textbooks, and courses on network theory. This approach generally focuses on microscopic models. As it applies to social media systems, sociology and computer science fields are useful areas of study.

Macroscopic epidemiology-based models for information (or rumor) spread are a hot topic of interest in papers and journals involving computational social systems. Books on traditional computational epidemiology are readily available and, paired with recent academic papers, should give one a good understanding of the material.

Data science, data collection, and data analysis are very common fields for researching local and online populations, events, and ideas. A rudimentary understanding of these topics will help any researcher, especially when doing case studies or trying to verify models. A programming foundation in a flexible and popular programming language such as Python is a great starting point, but there are several books, courses, and free online tutorials to aid in learning data science basics. For more advanced users, machine learning is quickly becoming an important tool in any analysis or prediction method that can utilize large amounts of data.

While reasonably simple control methods are touched on in this text, they only begin to scratch the surface of the discipline as a whole. More advanced methods such as optimal control offer personalized and practical alternatives, though they are often more rigorous from a mathematical and design perspective. A strong understanding of mathematics, engineering, or applied sciences is encouraged before delving too deeply into the control of complicated systems. Additionally, some programming experience is necessary for real-world (messy) data to which a control scheme is being applied. That said, learning control methods are very common in engineering courses of all types and there is a wealth of literature on the topic. The primary unique challenge is in intelligently applying control methods to social media systems.

16.3   The Future and Beyond

Where might we see socio-technical systems and information spread moving in the near future? How might things change? As technology advances so does our adoption of that technology and ultimately our culture. A key aspect of this culture change is how we spread ideas, desires, entertainment, political opinions, and more. As new models arise to address these new cultural information spread mediums, the current models will evolve or perhaps not even be needed in the future. The concepts behind them, however, may persist and lay the groundwork for future research and understanding.

Already there are virtual reality hangout events for movies, sports, and holidays where information is exchanged in an increasingly personal (yet physically distanced) way. Additionally, the role of automation and artificial intelligence is on the rise in nearly every aspect of our lives. Perhaps these technologies will also change the way we communicate and perceive information. In fact, we already see signs of this with targeted news stories and advertisements constructed by algorithms and powered by the massive amounts of data collected from each of us. Will misinformation, socio-cultural echo chambers, deep fake videos, and mass manipulation be common place or will we put in place some protections against them? Ultimately the answer is unclear. The only thing we can be sure of is that continued curiosity, exploration, modeling, testing, and implementing how information and ideas move throughout society will be at the forefront of the resolution to these questions.

Bibliography

[1] Stephen J Grove, Gregory M Pickett, and David N Laband. An empirical examination of factual information content among service advertisements. *Service Industries Journal*, 15(2):203–215, 1995.

[2] Eytan Bakshy, Solomon Messing, and Lada A Adamic. Exposure to ideologically diverse news and opinion on facebook. *Science*, 348(6239):1130–1132, 2015.

[3] Hendrik Bessembinder, Kalok Chan, and Paul J Seguin. An empirical examination of information, differences of opinion, and trading activity. *Journal of Financial Economics*, 40(1):105–134, 1996.

[4] Nicole A Cooke. *Fake news and alternative facts: Information literacy in a post-truth era*. American Library Association, 2018.

[5] Alban Galland, Serge Abiteboul, Amélie Marian, and Pierre Senellart. Corroborating information from disagreeing views. In *Proceedings of the third ACM international conference on Web search and data mining*, pages 131–140, 2010.

[6] Hafizh A Prasetya and Tsuyoshi Murata. A model of opinion and propagation structure polarization in social media. *Computational Social Networks*, 7(1):1–35, 2020.

[7] Margaret E Harrison. Collecting sensitive and contentious information. *Doing Development Research*, pages 115–129, 2006.

[8] Dietram A Scheufele and Nicole M Krause. Science audiences, misinformation, and fake news. *Proceedings of the National Academy of Sciences*, 116(16):7662–7669, 2019.

[9] World Health Organization. Coronavirus disease (COVID-19) advice for the public, 2020. `https://www.who.int/emergencies/diseases/novel-coronavirus-2019/advice-for-public`, Last accessed on 2020-6-1.

[10] United States Federal Emergency Management Agency. Coronavirus rumor control, 2020. `https://www.fema.gov/coronavirus/rumor-control`, Last accessed on 2020-6-1.

[11] Xiaochi Zhang. Internet rumors and intercultural ethics — a case study of panic-stricken rush for salt in China and iodine pill in America after Japanese earthquake and tsunami. *Studies in Literature and Language*, 4(2):13–16, 2012.

[12] Kerwin Swint. Founding fathers' dirty campaign. http://www.cnn.com/2008/LIVING/wayoflife/08/22/mf.campaign.slurs.slogans. Accessed: 2017-10-3.

[13] Don Reisinger. A look back at steve jobs and apple's "get a mac" ads. http://fortune.com/2016/12/09/apple-get-a-mac-ads/. Accessed: 2017-9-26.

[14] Caitlin Drummond and Baruch Fischhoff. Individuals with greater science literacy and education have more polarized beliefs on controversial science topics. *Proceedings of the National Academy of Sciences*, 114(36):9587–9592, 2017.

[15] Josh Constine. Whatsapp's first ads appear on Facebook and start convos with businesses. https://techcrunch.com/2017/09/08/whatsapp-ads/. Accessed: 2017-11-2.

[16] Aarti Shahani. From hate speech to fake news: The content crisis facing Mark Zuckerberg. http://fortune.com/2016/12/09/apple-get-a-mac-ads/. Accessed: 2017-11-6.

[17] Martyn Shuttleworth. History of the philosophy of science, Explorable. com. Sep 2009.

[18] Alina Bradford. Deductive reasoning vs. inductive reasoning. *Live Science*, 2017.

[19] Hillel Ofek. Why the arabic world turned away from science. *The New Atlantis*, JSTOR. pages 3–23, 2011.

[20] Robert A. Nisbet and Liah Greenfeld. Social science. Encyclopædia Britannica. October 16, 2020. https://www.britannica.com/topic/social-science Accessed: 2020-12-19.

[21] George C Homans. Social behavior as exchange. *American Journal of Sociology*, 63(6):597–606, 1958.

[22] Richard M Emerson. Social exchange theory. *Annual Review of Sociology*, 2(1):335–362, 1976.

[23] Marianna Sigala, Evangelos Christou, and Ulrike Gretzel. *Social media in travel, tourism and hospitality: Theory, practice and cases.* Ashgate Publishing, Ltd., 2012.

[24] Marshall McLuhan. Media hot and cold. *Understanding Media: The Extensions of Man*, MIT press. pages 22–32, 1994.

[25] Eytan Bakshy, Itamar Rosenn, Cameron Marlow, and Lada Adamic. The role of social networks in information diffusion. In *Proceedings of the 21st international conference on World Wide Web*, pages 519–528. ACM, 2012.

[26] Oren Tsur and Ari Rappoport. What's in a hashtag?: Content based prediction of the spread of ideas in microblogging communities. In *Proceedings of the fifth ACM international conference on Web search and data mining*, pages 643–652. ACM, 2012.

[27] Lilian Weng. *Information diffusion on online social networks*. PhD thesis, Indiana University, 2014.

[28] Kristina Lerman and Rumi Ghosh. Information contagion: An empirical study of the spread of news on Digg and Twitter social networks. *ICWSM*, 10:90–97, 2010.

[29] TrendStream. Global web index, 2010. https://www.globalwebindex.com/data

[30] Anjala S Krishen, Orie Berezan, Shaurya Agarwal, and Pushkin Kachroo. The generation of virtual needs: Recipes for satisfaction in social media networking. *Journal of Business Research*, 69(11):5248–5254, 2016.

[31] Stanley Wasserman and Katherine Faust. *Social network analysis: Methods and applications*, volume 8. Cambridge University Press, 1994.

[32] Charles Kadushin. *Understanding social networks: Theories, concepts, and findings*. OUP USA, 2012.

[33] Paul F Lazarsfeld, Robert K Merton, et al. Friendship as a social process: A substantive and methodological analysis. *Freedom and Control in Modern Society*, 18(1):18–66, 1954.

[34] Engin Bozdag. Bias in algorithmic filtering and personalization. *Ethics and Information Technology*, 15(3):209–227, 2013.

[35] Dominic DiFranzo and Kristine Gloria-Garcia. Filter bubbles and fake news. *XRDS: Crossroads, The ACM Magazine for Students*, 23(3):32–35, 2017.

[36] Megan Brenan. *Americans' Trust in Mass Media Edges Down to 41 Percent*, 2019. https://news.gallup.com/poll/267047/americans-trust-mass-media-edges-down.aspx. Accessed on 2020-5-1.

[37] Georg Simmel. *The sociology of georg simmel*, volume 92892. Simon and Schuster, 1950.

[38] Fritz Heider. Attitudes and cognitive organization. *The Journal of Psychology*, 21(1):107–112, 1946.

[39] Glenn Lawyer. Understanding the influence of all nodes in a network. *Scientific Reports*, 5:8665, 2015.

[40] Ronald S Burt. *Structural holes: The social structure of competition.* Harvard University Press, 2009.

[41] Alex Bavelas. Communication patterns in task-oriented groups. *The Journal of the Acoustical Society of America*, 22(6):725–730, 1950.

[42] Jérémie Bouttier, Philippe Di Francesco, and Emmanuel Guitter. Geodesic distance in planar graphs. *Nuclear Physics B*, 663(3):535–567, 2003.

[43] Sriram Pemmaraju and Steven Skiena. *Computational discrete mathematics: Combinatorics and graph theory with Mathematica®.* Cambridge University Press, 2003.

[44] Edsger W Dijkstra. A note on two problems in connexion with graphs. *Numerische Mathematik*, 1(1):269–271, 1959.

[45] Duncan J Watts and Steven H Strogatz. Collective dynamics of "small-world"networks. *Nature*, 393(6684):440–442, 1998.

[46] H Cooley Charles. Social organization: A study of the larger mind. *Charles Scribner's Sons*, New York, page 23, 1909.

[47] Lada A Adamic and Natalie Glance. The political blogosphere and the 2004 us election: Divided they blog. In *Proceedings of the 3rd international workshop on Link discovery*, pages 36–43. ACM, 2005.

[48] Caleb Jones. Visualizing polarization in political blogs, Oct 2014. http://allthingsgraphed.com/2014/10/09/visualizing-political-polarization/

[49] Song Yang, Franziska B Keller, and Lu Zheng. *Social network analysis: Methods and examples.* Sage Publications, 2016.

[50] Dimitar Nikolov, Diego FM Oliveira, Alessandro Flammini, and Filippo Menczer. Measuring online social bubbles. *PeerJ Computer Science*, 1:e38, 2015.

[51] Pankaj Maheshwari, Romesh Khaddar, Pushkin Kachroo, and Alexander Paz. Dynamic modeling of performance indices for planning of sustainable transportation systems. *Networks and Spatial Economics*, 16(1):371–393, 2016.

[52] Shaurya Agarwal, Pushkin Kachroo, Sergio Contreras, and Shankar Sastry. Feedback-coordinated ramp control of consecutive on-ramps using distributed modeling and godunov-based satisfiable allocation. *IEEE Transactions on Intelligent Transportation Systems*, 16(5):2384–2392, 2015.

[53] Pushkin Kachroo, Shaurya Agarwal, and Shankar Sastry. Inverse problem for non viscous mean field control: Example from traffic. *IEEE Transactions on Automatic Control*, 61(11):3412–3421, 2016.

[54] Pushkin Kachroo, Shaurya Agarwal, Benedetto Piccoli, and Kaan Ozbay. Multi-scale modeling and control architecture for v2x enabled traffic streams. *IEEE Transactions on Vehicular Technology*, 66(6): 4616–4626, 2017.

[55] MJ Lighthill and GB Whitham. On kinematic waves. i: flow movement in long rivers. ii: a theory of traffic on long crowded roods. In *Proceeding of the Royal Society*, number A229, pages 281–345, 1955.

[56] PI Richards. Shockwaves on the highway. *Operationa Research*, 4:42–51, 1956.

[57] BD Greenshields. A study in highway capacity. *Highway Research Board*, 14:458, 1935.

[58] Shaurya Agarwal, Pushkin Kachroo, and Sergio Contreras. A dynamic network modeling-based approach for traffic observability problem. *IEEE Transactions on Intelligent Transportation Systems*, 17(4):1168–1178, 2015.

[59] Herbert W Hethcote. Three basic epidemiological models. In *Applied mathematical ecology*, pages 119–144. Springer, 1989.

[60] Gypsyamber D'Souza and David Dowdy. What is herd immunity and how can we achieve it with COVID-19?, Apr 2020. https://www.jhsph.edu/covid-19/articles/achieving-herd-immunity-with-covid19.html

[61] Morgan Krakow. A tourist infected with measles visited disneyland and other southern california hot spots in mid-august, Aug 2019. https://www.washingtonpost.com/health/2019/08/24/tourist-infected-with-measles-visited-disneyland-other-southern-california-hotspots-mid-august/

[62] Sergio Contreras, Pushkin Kachroo, and Shaurya Agarwal. Observability and sensor placement problem on highway segments: A traffic dynamics-based approach. *IEEE Transactions on Intelligent Transportation Systems*, 17(3):848–858, 2016.

[63] Michael Muhlmeyer, Shaurya Agarwal, and Jiheng Huang. Modeling social contagion and information diffusion in complex socio-technical systems. *IEEE Systems Journal*, 2020.

[64] ML Vidale and HB Wolfe. An operations-research study of sales response to advertising. *Operations Research*, 5(3):370–381, 1957.

[65] Albert Einstein. *Investigations on the theory of the Brownian movement.* Courier Corporation, 1956.

[66] Pratik Verma, Hongtao Yang, Pushkin Kachroo, and Shaurya Agarwal. Modeling and estimation of the vehicle-miles traveled tax rate using stochastic differential equations. *IEEE Transactions on Systems, Man, and Cybernetics: Systems*, 46(6):818–828, 2016.

[67] Linda JS Allen, Fred Brauer, Pauline Van den Driessche, and Jianhong Wu. *Mathematical epidemiology*, volume 1945. Springer, 2008.

[68] Jayson DeMers. The top 10 benefits of social media marketing. https://www.forbes.com/sites/jaysondemers/2014/08/11/the-top-10-benefits-of-social-media-marketing/786102ba1f80. Accessed: 2018-6-21.

[69] Michael Muhlmeyer, Jiheng Huang, and Shaurya Agarwal. Event triggered social media chatter: A new modeling framework. *IEEE Transactions on Computational Social Systems*, 6(2):197–207, 2019.

[70] S Goyal. Advertising on social media. *Scientific Journal of Pure and Applied Sciences*, 2(5):220–223, 2013.

[71] Prasad A Naik et al. Marketing dynamics: A primer on estimation and control. *Foundations and Trends® in Marketing*, 9(3):175–266, 2015.

[72] Gerhard Sorger. Competitive dynamic advertising: A modification of the case game. *Journal of Economic Dynamics and Control*, 13(1):55–80, 1989.

[73] Chuang Liu, Xiu-Xiu Zhan, Zi-Ke Zhang, Gui-Quan Sun, and Pak Ming Hui. How events determine spreading patterns: Information transmission via internal and external influences on social networks. *New Journal of Physics*, 17(11):113045, 2015.

[74] Shan-Hung Wu, Man-Ju Chou, Chun-Hsiung Tseng, Yuh-Jye Lee, and Kuan-Ta Chen. Detecting in situ identity fraud on social network services: A case study with facebook. *IEEE Systems Journal*, 11(4):2432–2443, 2015.

[75] Yunji Liang, Xingshe Zhou, Daniel Dajun Zeng, Bin Guo, Xiaolong Zheng, and Zhiwen Yu. An integrated approach of sensing tobacco-oriented activities in online participatory media. *IEEE Systems Journal*, 10(3):1193–1202, 2014.

[76] Hajar Rehioui and Abdellah Idrissi. New clustering algorithms for twitter sentiment analysis. *IEEE Systems Journal*, 14(1): 530–537, 2019.

[77] Investigative assistance for violent crimes act of 2012. Pub. L. No. 112–256 § 2, 126 Stat. 2435. 2013. [Online]. `https://www.congress.gov/112/plaws/publ265/PLAW-112publ265.pdf`. Accessed: 2020-7-15

[78] Ari Schulman. How not to cover mass shootings. `https://www.wsj.com/articles/how-not-to-cover-mass-shootings-1510939088/`, Accessed on 2018-5-25.

[79] Mass shooting tracker. `https://www.massshootingtracker.org/data/` 2017. Accessed: 2018-5-03.

[80] Jean-Jacques E Slotine, Weiping Li, et al. *Applied nonlinear control*, volume 199. Prentice hall Englewood Cliffs, NJ, 1991.

[81] M Peifer and J Timmer. Parameter estimation in ordinary differential equations for biochemical processes using the method of multiple shooting. *IET Systems Biology*, 1(2):78–88, 2007.

[82] Zhengfeng Li, Michael R Osborne, and Tania Prvan. Parameter estimation of ordinary differential equations. *IMA Journal of Numerical Analysis*, 25(2):264–285, 2005.

[83] Sophie Gilbert. The movement of #MeToo: How a hashtag got its power. `https://www.theatlantic.com/entertainment/archive/2017/10/the-movement-of-metoo/542979/`, Accessed on 2018-6-22.

[84] KT Hawbaker Christen A Johnson. #MeToo: A timeline of events, 2018 `https://www.chicagotribune.com/lifestyles/ct-me-too-timeline-20171208-htmlstory.html`, Accessed on 2018-11-7.

[85] Ahmet Taspinar. Scrape Twitter for tweets. `https://github.com/taspinar/twitterscraper`, 2018. Accessed on 2018-3-19.

[86] Megan McClusky. Gillette makes waves with controversial new ad highlighting toxic masculinity. `http://time.com/5503156/gillette-razors-toxic-masculinity/`, Accessed on 2019-2-8.

[87] Dusty Baxter-Wright. 7 viral things the internet has been completely divided over. `https://www.cosmopolitan.com/uk/entertainment/a20726869/viral-things-divided-the-internet/`, Accessed on 2018-2-5.

[88] Richard C Dorf and Robert H Bishop. *Modern control systems*. Pearson, 2011.

[89] Shaurya Agarwal and Pushkin Kachroo. Controllability and observability analysis for intelligent transportation systems. *Transportation in Developing Economies*, 5(1):2, 2019.

[90] Karl Johan Åström, Tore Hägglund, and Karl J Astrom. *Advanced PID control*, volume 461. ISA-The Instrumentation, Systems, and Automation Society Research Triangle?, 2006.

[91] Michael A Johnson and Mohammad H Moradi. *PID control*. Springer, 2005.

[92] Hector J Sussmann and Jan C Willems. 300 years of optimal control: From the brachystochrone to the maximum principle. *IEEE Control Systems Magazine*, 17(3):32–44, 1997.

[93] Donald E Kirk. *Optimal control theory: An introduction*. Courier Corporation, 2012.

[94] Pankaj Maheshwari, Pushkin Kachroo, Alexander Paz, and Romesh Khaddar. Development of control models for the planning of sustainable transportation systems. *Transportation Research Part C: Emerging Technologies*, 55:474–485, 2015.

[95] Kundan Kandhway and Joy Kuri. Campaigning in heterogeneous social networks: Optimal control of SI information epidemics. *IEEE/ACM Transactions on Networking*, 24(1):383–396, 2016.

[96] Kundan Kandhway and Joy Kuri. Optimal control of information epidemics modeled as maki thompson rumors. *Communications in Nonlinear Science and Numerical Simulation*, 19(12):4135–4147, 2014.

[97] Pushkin Kachroo, Saumya Gupta, Shaurya Agarwal, and Kaan Ozbay. Optimal control for congestion pricing: Theory, simulation, and evaluation. *IEEE Transactions on Intelligent Transportation Systems*, 18(5):1234–1240, 2017.

Index